Advances in Bioceramics and Biocomposites II

Advances in Bioceramics and Biocomposites II

A Collection of Papers Presented at the 30th International Conference on Advanced Ceramics and Composites January 22–27, 2006, Cocoa Beach, Florida

Editor
Mineo Mizuno

General Editors
Andrew Wereszczak
Edgar Lara-Curzio

A JOHN WILEY & SONS, INC., PUBLICATION

Published by John Wiley & Sons, Inc., Hoboken, New Jersey
Published simultaneously in Canada.

Library of Congress Cataloging-in-Publication Data is available.

ISBN-13 978-0-470-08056-6
ISBN-10 0-470-08056-6

10 9 8 7 6 5 4 3 2 1

Contents

Preface

Bones and teeth are composed of inorganic (calcium phosphate) and organic (proteins) materials. They are ultimate composites, being skillfully tailored to show both structural and bioactive functions. The role of biomaterials is important because life expectancy continues to increase in our society along with the needs of this segment of the population.

It was a timely fashion that a "Bioceramics and Biocomposites" session started in 2002 in the 26th International Conference on Advanced Ceramics and Composites, and growing interest in this topic resulted in the organization of a full symposium in 2005. At present, bioceramics are undoubtedly recognized to be one of most important materials in order to overcome problems on the aging society in the near future. The use of ceramics in biological environments and biomedical applications is of increasing importance, as is the understanding of how biology works with minerals to develop strong materials.

These proceedings contain papers that were presented at the Bioceramics and Biocomposites symposium held at the 30th International Conference and Exposition on Advanced Ceramics and Composites, Cocoa Beach, Florida, January 22–27, 2006. Authors from 12 different countries, representing academia, national laboratories, industries, and government agencies, presented a total of 43 papers at this symposium.

This symposium covered topics such as processing of biomaterials, orthopaedic replacements and performance issue of biomaterials, biomimetics and biocomposites, drug delivery and dental ceramics.

The symposium organizers would like to thank all the participants in the symposium and the staff at The American Ceramic Society. The symposium organizers hope that this symposium will inspire the development of better biomaterials to contribute to improvements in the quality of life.

MINEO MIZUNO
JIAN KU SHANG
RICHARD RUSIN
WALTRAUD KRIVEN

Introduction

This book is one of seven issues that comprise Volume 27 of the Ceramic Engineering & Science Proceedings (CESP). This volume contains manuscripts that were presented at the 30th International Conference on Advanced Ceramic and Composites (ICACC) held in Cocoa Beach, Florida January 22–27, 2006. This meeting, which has become the premier international forum for the dissemination of information pertaining to the processing, properties and behavior of structural and multifunctional ceramics and composites, emerging ceramic technologies and applications of engineering ceramics, was organized by the Engineering Ceramics Division (ECD) of The American Ceramic Society (ACerS) in collaboration with ACerS Nuclear and Environmental Technology Division (NETD).

The 30th ICACC attracted more than 900 scientists and engineers from 27 countries and was organized into the following seven symposia:

- Mechanical Properties and Performance of Engineering Ceramics and Composites
- Advanced Ceramic Coatings for Structural, Environmental and Functional Applications
- 3rd International Symposium for Solid Oxide Fuel Cells
- Ceramics in Nuclear and Alternative Energy Applications
- Bioceramics and Biocomposites
- Topics in Ceramic Armor
- Synthesis and Processing of Nanostructured Materials

The organization of the Cocoa Beach meeting and the publication of these proceedings were possible thanks to the tireless dedication of many ECD and NETD volunteers and the professional staff of The American Ceramic Society.

ANDREW A. WERESZCZAK
EDGAR LARA-CURZIO
General Editors

Oak Ridge, TN (July 2006)

In Vitro Evaluation

INITIAL IN VITRO INTERACTION OF HUMAN OSTEOBLASTS WITH NANOSTRUCTURED HYDROXYAPATITE (NHA)

XINGYUAN GUO[1], JULIE GOUGH[1], PING XIAO[1]
1. Manchester Materials Science Centre, School of Materials, The University of Manchester, Grosvenor street, Manchester, UK, M1 7AZ

JING LIU[2] ZHIJIAN SHEN[2]
2. Department of Inorganic Chemistry, Arrhenius Laboratory, Stockholm University, S-10691, Stockholm, Sweden

ABSTRACT

Nanostructured hydroxyapatite (NHA) was fabricated by Spark Plasma Sintering (SPS), while microstructured hydroxyapatite (MHA) by conventional method. Human Osteoblasts were cultured on both NHA and MHA and the cell attachment, proliferation and mineralisation were evaluated. After 90 min incubation the cell density on NHA surface is significantly higher than that of MHA and glass control, whereas average cell area of a spread cell is significantly lower on NHA surface compared to MHA and glass control after 4 h incubation. Mineralisation of matrix has been determined after 14 days culture by using alizarin red assay combined with cetylpyridinium chloride (CPC) extraction. NHA shows significant enhancement ($p<0.05$) in mineralisation compared to MHA. Results from this study suggest that NHA is a much better candidate for clinical uses in terms of bioactivity.

1. INTRODUCTION

Interaction between cells and implanted materials depends on the physical and chemical characteristics of materials and particularly on its chemical composition, particle size and surface properties, which include their topography, roughness, surface energy, hydrophilicity and hydrophobicity[1-3]. Such characteristics determine how biological molecules will adsorb to the surface. Specifically, maximum vitronectin (a protein contained in serum that is known to mediate osteoblast adhesion[4]), fibronectin and albumin adsorption was noted on hydrophilic surface with high surface roughness and /or energies, such selected protein has been identified to mediate adhesion of specific anchorage-dependent cells (such as osteoblasts, fibroblasts, and endothelial cells) on substrate surfaces[5]. Therefore, surface properties will affect cell adhesion, attachment on implanted materials in the first phase of cell/material interactions and thus further influence the cell's capacity to proliferate and to differentiate on contact with the implant[2, 6]. Design of materials with improved physical and chemical properties could enhance cell response to biomaterial implants and further extend their lifetime and, therefore, decrease the rate of revision surgery.

It has been reported that surface properties (such as surface area, charge, and topography) depend on the grain size of a material.[5] In this respect, nanostructured materials possess higher surface area with increased portions of surface defects and grain-boundaries.[7] Meanwhile, hydroxyapatite has been considered as a good candidate for designing hard tissue implants due to its excellent biological properties such as non-toxicity, lack of inflammatory response and immunological reactions, and is able to intimately bond to new bone[8, 9]. Consequently, It is extremely attractive to explore if and how nanostructured hydroxyapatite with enhanced surface properties (such as increased surface area and charge, as well as ability to alter adsorption of chemical species) could be used to promote cell response and bonding of juxtaposed bone to an orthopaedic or dental implants composed of nanophase ceramics. However, only a few reports have been published till now. Webster has reported that osteoblasts adhesion and osteoblasts proliferation was significantly greater on nanophase alumina, titania, and hydroxyapatite than on conventional formulations of the same ceramic after 3 and 5 days and more importantly, compared to conventional ceramics, synthesis of alkaline phosphatase and deposition of calcium containing mineral was significantly greater by osteoblasts cultured on nanophase than conventional ceramics after 21 days and 28 days.[5, 10, 11] In this article, a primary human osteoblast cell model has been used to study the cellular response to nanostructured HA (NHA) compared with microstructured HA (MHA).

2 EXPERIMENTS

2.1 Materials

HA powder has been prepared by precipitation at room temperature using $Ca(OH)_2$ and H_3PO_4 as starting materials[12]. Such powder has been characterized using XRD and FTIR as a hydroxyapatite phase with low crystallinity and with incorporation of carbonate ions. The shape of the HA crystals are acicular according to TEM images. Crystallite size calculated according to XRD results is about 20-40 nm, while the particle size measured by Mastersizer microplus (Malvern Ltd, UK) is several microns. All nanostructured HA (NHA) samples used for cell culture were sintered at 900°C for 3 min by the SPS process with a heating rate of $100^{o}C \cdot min^{-1}$ and a pressure of 50 MPa at vacuum atmosphere. All microstructured HA (MHA) compacts were sintered by conventional method at 1200°C for 2 h with a heating rate of $5^{o}C \cdot min^{-1}$ in air.

2.2 Surface characterisation

The sintered compacts were polished using Silica colloid (0.06 μm) and chemically etched in 18.0 mM HCl solution to reveal the grain boundaries, and then microstructural observations of sintered ceramics were conducted using a high-resolution scanning electron microscope (FESEM-XL30, Philips). Topography and surface roughness of NHA and MHA has been evaluated by Atomic force microscopy (AFM). Five measurements were made on each sample with a scanning area of 5 x 5 μm. The classical mean surface roughness parameter R_a has been used to characterize surface roughness. Aqueous wettability of nanostructured HA and microstructured HA has been analysed by contact angle measurements on polished samples. Measurements were run in triplicate per sample and repeated at three different times.

2.3 Cell morphology

HOBs in complete Dulbecco's modified Eagles medium (DMEM) containing 10% foetal bovine serum (FBS), 1% antibiotics and 0.85mM ascorbic acid-2 phosphate were seeded onto microstructured HA (MHA) and nanostructured HA (NHA) discs at a density of $4x10^4$ cells/cm^2, then incubated at 37°C in a humidified incubator with 5% CO_2 for 90 min, 4 h, and 24 h. Glass coverslips purchased from Chance Glass Ltd were used as control materials. At the pre-determined time points, samples were rinsed in phosphate buffered saline (PBS) to remove any non-adherent cells and medium. The remaining cells were fixed in 1.5% glutaraldehyde for 30 min at 4°C, then dehydrated through a series of ethanol concentrations (50%, 70%, 90%, and 100%) and dried using hexamethyldisilazane (HMDS). Once dry, the samples were coated with gold and examined in a JEOL JSM-840 Scanning Electron Microscope at 10kV[13, 14].

2.4 Cell proliferation

Cells were harvested with 0.1% trypsin-EDTA solution in phosphate-buffered saline (PBS, pH 7.4) from cell culture flasks and were resuspended in culture medium, then were seeded at a concentration of $4x10^4$ cells/cm^2 onto disks of MHA and NHA. Tissue culture polystyrene was used as a control material. Cells were left to grow on the disks for 1, 3, 7 days in a 37°C incubator with 5% CO_2. At the pre-determined time points, each of the disks was transferred to new wells in a new 24-well plate and 1.5 ml medium were added to each disk. 150 µl of freshly prepared 5 mg/ml 3-(4,5-dimethythiazolyl-2)-2,5- diphenyltetrazolium bromide (MTT) were added to each well containing the disks. The plates were placed in an incubator at 37°C for 3 h. Afterwards, the supernatant of each well were removed and acidified isopropanol (0.04 M HCl in isopropanol) was added to all wells and mixed thoroughly to dissolve the dark-blue crystals. After all crystals were dissolved, the plates were read on a Microplate reader (Ascent) with a test wavelength of 540 nm against a reference wavelength of 620 nm[15].

2.5 Alamar Blue™ assay

Cells were harvested with 0.1% trypsin-EDTA solution in phosphate-buffered saline (PBS, pH 7.4) from cell culture flasks and were resuspended in culture medium, then were seeded at a concentration of $4x10^4$ cells/cm^2 onto disks of MHA and NHA. Tissue culture polystyrene was used as a control material. After 1 day incubation in a 37°C incubator with 5% CO_2, samples were taken out into new wells and 1.25 µg/ml Alamar Blue™ were added into each well, then incubated for 3 h. The fluorescence was measured using a FLUOstar OPTIMA (BMG LABTECH) plate reader at 530ex-590em nm wavelengths. The suspension containing Alamar Blue™ was removed from sample wells and new medium added to incubate further[16, 17].

2.6 Mineralisation

Ability of cells to produce mineralised matrix is essential with regard to development of materials for bone regeneration. Whether mineralisation of matrix occurred was determined using alizarin red-S (AR-S) assay combined with cetylpyridinium chloride (CPC) extraction[18, 19]. Alizarin red is a dye which binds selectively to calcium salts and is widely used for calcium mineral histochemistry. AR-S binds ~2 mol of Ca^{2+} /mol of dye in solution. Briefly, Cells were seeded onto disks of MHA and NHA at a concentration of $4x10^4$ cells/cm^2 by using medium, which was supplemented with 10 mM β-glycerophosphate and 100 nM dexamethasone. Tissue culture polystyrene was used as a control material. At the pre-determined time points (7,14 days), samples were briefly rinsed with PBS followed by fixation (ice-cold 70% ethanol, 1 h). Samples then were rinsed with nanopure water and stained for 10 min with 40 mM AR-S, pH 4.2, at room temperature. Afterwards, samples were rinsed five times with water followed by a 15 min wash with PBS to reduce non-specific AR-S stain. Stained cultures were destained by using 10% (w/v) cetylpyridinium chloride (CPC) in 10 nm sodium phosphate, pH 7.0, for 15 min at room temperature. After destain, the plates were read on a Microplate reader (Ascent) with a test wavelength of 540 nm against a reference wavelength of 620 nm. Furthermore, for determination of the matrix mineralisation directly on HA samples, two groups of samples were used. Group1 was cultured with HOBs, and Group 2 without HOBs as background. To calculate the mineral content of the extracellular matrix, group 2 values were subtracted from group 1.

2.7 Statistical analysis

All samples were run in triplicate and repeated three times. Statistical analysis was performed using student's t-test, and a p value of <0.05 was determined to represent a significant difference.

3 RESULTS

Fig.1 shows the scanning electronic microstructure of NHA and MHA. Dense compact consisting of equiaxed grains with an average grain size of ~90 nm was obtained by SPS (as shown in Fig. 1A), whereas the grain size is about 2 μm when sintering by conventional method at 1200°C as shown in Fig. 1B. The relative density of NHA is 100%, whereas 95% of MHA measured by Archimedes method with deionized water as the immersion medium. After sintering, NHA was characterized as pure HA with nanostructured feature, while MHA was characterized with slightly decomposition to tetracalcium phosphate (results not shown in this article).

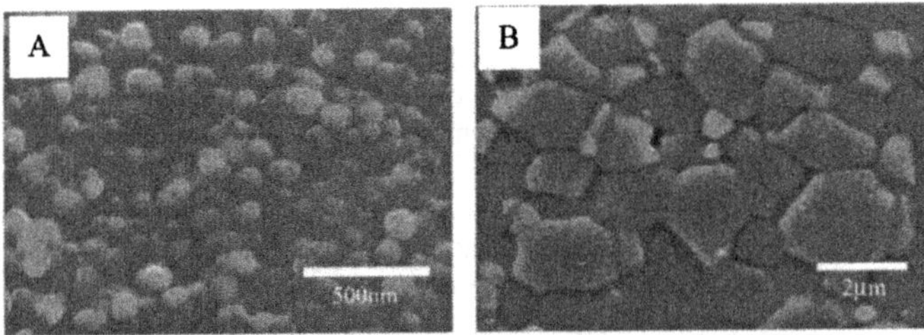

Fig.1 SEM morphologies of (A) NHA and (B) MHA

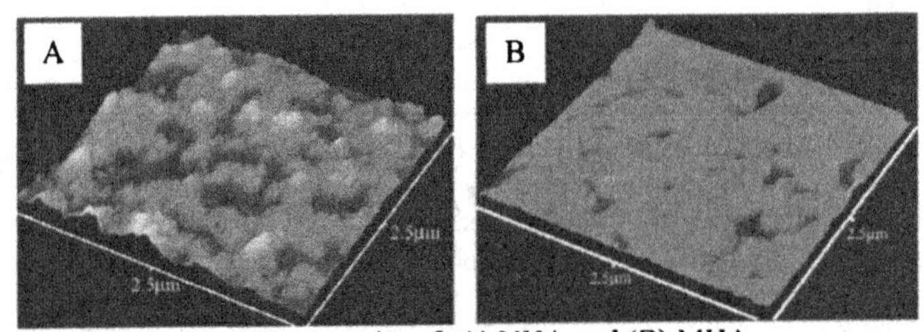

Fig.2 Topography of (A) NHA and (B) MHA

According to contact angle measurement, aqueous contact angle on NHA is 30°, whereas 42° on MHA, which was significantly ($p<0.05$) higher than that on NHA. The decrease in contact angle corresponds to an increase in surface aqueous Wettability and, thus, an increase in hydrophilicity and surface reactivity for NHA. Topography measured by Atomic force microscopy (AFM) presents in Fig.2 shows that surface of MHA is much smoother than that of NHA. Surface roughness calculated from AFM data provided evidence that NHA possesses significantly higher roughness (37.2 ± 5.5 nm) compared to MHA (12.1±3.0 nm).

Fig. 3 shows the SEM morphologies of human osteoblasts cultured on NHA and MHA after 90 min, 4 h, and 24 h with glass as control material. Different morphologies have been observed on various surfaces. At 90 min most cells present a spherical, spiky appearance on both HA surfaces, whereas more spread cells are observed on glass. Abundant filopodia were visible at the edge of cells on both HA surfaces, in particular on the NHA surface. Those spiky and more three-dimensional morphologies of cells cultured on HA surfaces might be attributed to surface

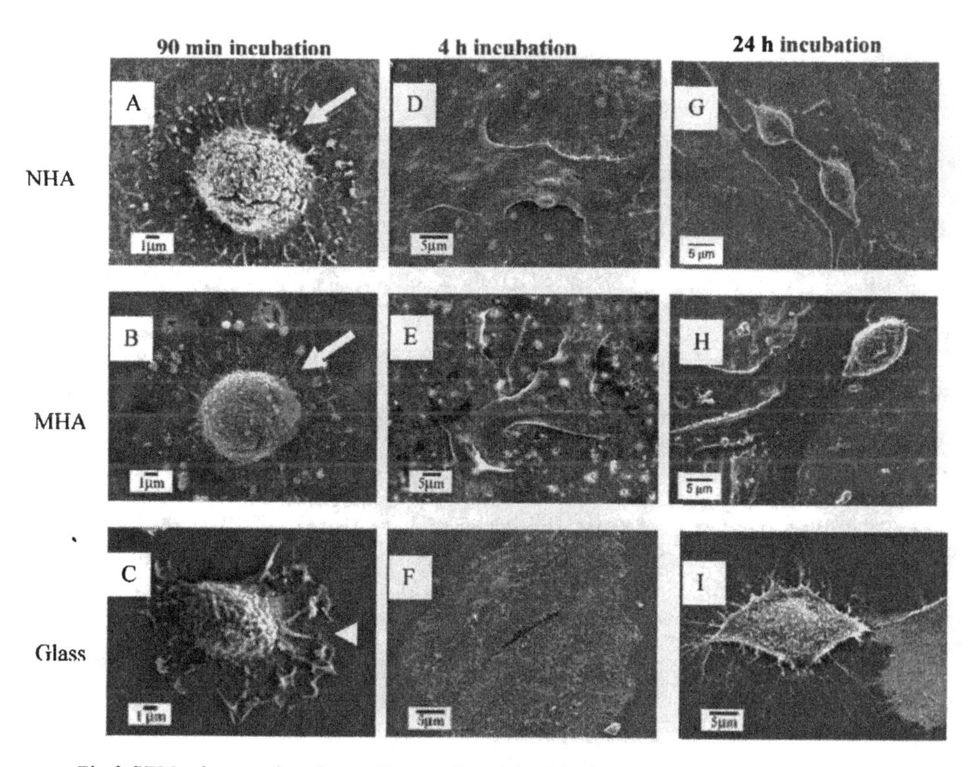

Fig.3 SEM micrographs of osteoblasts cultured for 90 min (A-C), 4 h (D-F), 24 h (G-I) on the surface of NHA, MHA and glass control respectively. Arrow indicates filopodia, arrowhead indicates ruffles.

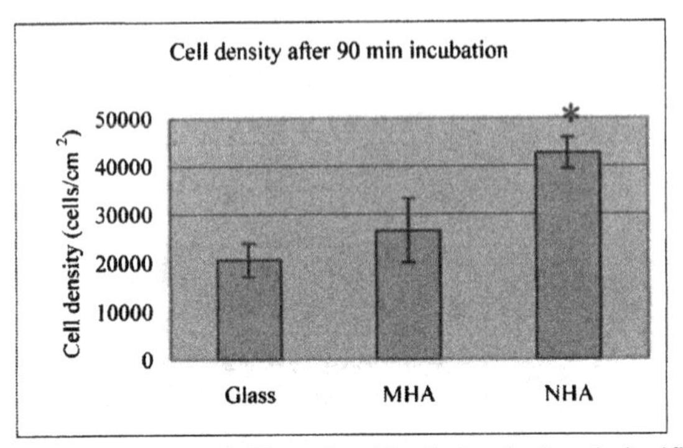

Fig. 4 Cell density on NHA, MHA and Glass after 90 min incubation, * significantly higher than MHA and Glass ($p<0.05$, $n=5$)

roughness.[14] After 4 h incubation, cells spread on all surfaces with the edge of the cells becoming very thin. At 24 h, cells became thicker at the center of nucleus, and mitosis was evident on some surfaces.

The cell density on both HA samples and Glass sample after 90 min incubation has been measured by counting cells in 5 randomly selected areas under SEM with a magnification of 200, results of which is shown in Fig. 4. The cell density on NHA is significantly higher compared to that on MHA and glass, which suggested that nanostructured surface promote osteoblast adhesion
in the first stage. The average areas of spread cells on various samples after 4 h incubation were measured by image analysis software (UTHSCSA Image Tool) from 20 randomly selected cells, which is shown in Fig. 5. The results indicated that cell spread significantly widely on both glass and microstructured HA compared to nanostructured HA.

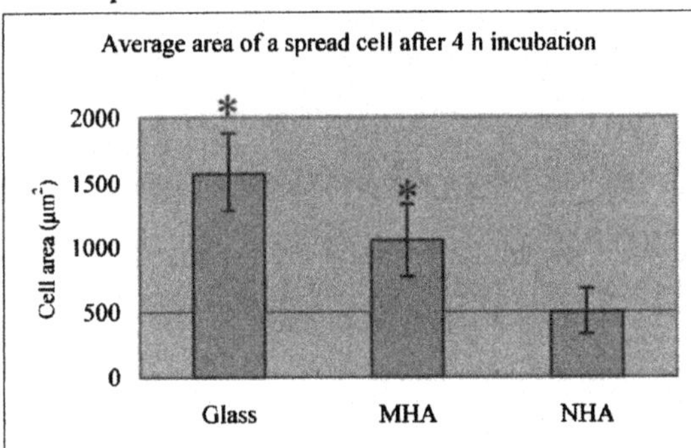

Fig. 5 The average area for a spread cell on various samples after 4 h incubation, 20 cells were measured from 5 randomly selected areas of each sample, and average area was calculated from the measurements. * Significant from NHA ($p<0.05$, $n=20$)

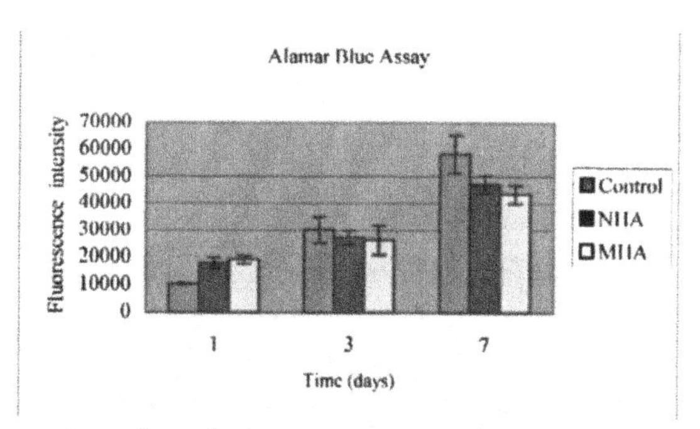

Fig 6 Cell viability on the surface of microstructure HA and nanostructured HA after 1,3,7 days incubation by Alamar Blue assay, control is tissue culture polystyrene

In order to study the long-term cell response, cell proliferation on various samples was measured by Alamar blue assay and MTT assay, which are presented at Fig. 6 and Fig.7, respectively. Both figures showed a similar trend that cell proliferation increased significantly ($p<0.05$, $n=3$) on both NHA and MHA from 1 day to 7 days. According to Alamar blue assay results, there was no significantly difference between each sample irrespective culture time. However, according to MTT results, cell proliferation was significantly higher on NHA compared to MHA ($p<0.05$, $n=3$) after 7 days incubation, though there was no significant difference between both surfaces after 1 day and 3 days incubation.

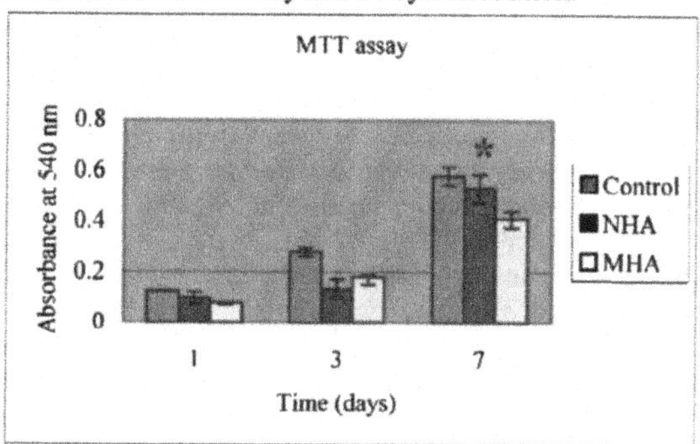

Fig.7 Osteoblasts proliferation on the surface of nanostructured HA and microstructured HA after 1,3,7days in culture. * Significant from microstructured HA, ($p<0.05$, $n=3$), control is tissue culture polystyrene

The ability of calcium-containing mineral deposition in extracellular matrix is essential for bone regeneration. The amount of calcium-containing mineral deposited onto the matrix has been quantified by alizarin red stain followed by CPC extraction, results of which are shown in Fig. 8. From 7 days to 14 days incubation, the amount of calcium-containing mineral on NHA surfaces and control increased significantly ($p<0.05$), but increase slowly on MHA surface.

Furthermore, the amount of calcium-containing mineral on NHA was significantly higher ($p<0.05$, n=3) compared to MHA after 14 days incubation.

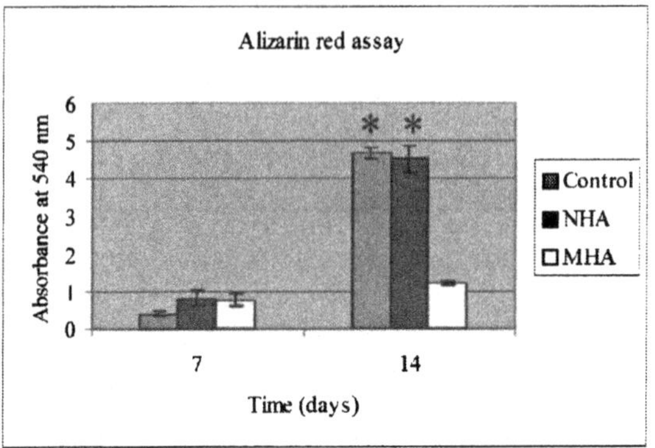

Fig.8 Quantification of alizarin red stain via extraction with 10% CPC in 10mM phosphate buffer on the surface of nanostructured HA and microstructured HA after 14 days in culture. * Significant from microstructure HA ($p<0.05$, n=3), control is tissue culture polystyrene

4. DISCUSSION

Nanostructured ceramics with grain size less than 100 nm have been widely studied these years due to enhanced magnetic, catalytic, electrical, and optical properties when compared to conventional formulations of the same materials[5]. Nanostructured materials also provide the capability for specific interactions with proteins, DNA, viruses, and other nanoscales biological structures. Highly specific interactions between these components and nanostructured materials can provide unique biological functionalities not seen with conventional microstructured materials[20]. In the present work, we examined the effect of grain size (nano-size, and micro-size) of HA on cellular adhesion, proliferation, and matrix mineralisation. The cell culture system used was human osteoblast.

Cells in contact with a surface will firstly attach, adhere and spread. This first phase of cell adhesion and spreading is known to affect the long-term phenotype of anchorage dependent cells[2, 14]. Osteoblasts, for example, are well documented in the literature, and it can be seen that on some surfaces they appear extremely flattened and on other surfaces they have a more three-dimensional morphology[21]. In this study, the cell morphology for both NHA and MHA as shown by SEM is three-dimensional, while quite flat for glass control. Cells on NHA surface presented more 'spiky' or stellate appearance than that on MHA surface, since more filopodia or microspikes have been observed on NHA surface. The filopodia or microspikes of cells are used in sensing the substrate. Microspikes of neurons extend over significant distances to determine areas suitable for attachment[22]. Therefore, the roughened surface in this study, i.e. NHA surface appears a greater surface area of the cell can adhere. Gough et.al[14] and Dalby et al[23] have observed similar morphologies of osteoblasts cultured on 45S5 bioactive glass and HAPEX™ (a composite of hydroxyapatite and polyethylene) with various roughened topography, respectively.

Cell adhesion was also quantified in this study by counting the cell density on various surfaces after 90 min incubation, the results of which show a significant increase of cell adhesion on NHA surface compared to MHA surface and Glass control. Such increased cell adhesion is in agreement with the more spiky morphology of osteoblasts on NHA surface. It has been previously found that the cell adhesion and spreading are influenced by the physico-chemical characteristics of the underlying solid surface, such as surface free energy of the substrate, surface charge, and surface topography of the biomaterials[24]. In addition, Thomos Webster et.al also found that osteoblast adhesion increased with grain size of ceramics decreasing to nano scale[11]. A possible explanation for such phenomena they suggested could be directly related to the greater surface area exhibited by NHA. The nanostructured surface might promote interaction (such as adsorption, configuration, bioactivity, etc.) of select serum protein(s), which subsequently, enhance osteoblast adhesion[11]. Proteins mediate adhesion of anchorage-dependant cells, and thus influence subsequent cellular functions (such as cell proliferation, matrix mineralisation etc.). The mechanism of protein interaction with nanostructured ceramic is not clear and needs to be further investigated.

Furthermore the average area of spread cell on NHA is significantly lower than that on MHA and Glass control materials. Similar results have been reported by Thomas Webster et.al.[10], which he suggested might be attributed to the motility of cells. The dynamics of cell migration require continuous forming and breaking of focal contacts at the proximal and distal sides of the cell, respectively. Cell migration is inhibited on substrate surfaces that promote cell adhesion; in other words, contractile forces necessary for cell migration cannot overcome the strength of cell contact points formed on highly adhesive substrate surfaces[25, 26]. Surface occupancy experiments have been utilized as an index of cell population motility[27]. The increased cell adhesion couples with decreased cell motility, as well as enhanced proliferation and matrix mineralisation has been observed on surface (for example, borosilicate glass and titanium) modified with immobilized peptide sequences (such as arginine-glycine-aspartic acid-serine (RGDS) and lysine-arginine-serine-arginine (KRSR)[28-30]) contained in extracellular matrix protein such as vitronectin and collagen.

Matrix mineralisation plays a critical role for the bonding (i.e. osseointegration) between biomaterials and natural bone and bone remodelling. Mineralisation on NHA was significantly greater ($p<0.05$) than that on MHA after 14 days culture in vitro, which indicates that NHA may greatly enhance osseointegration compared to microstructured HA, and subsequently promote the life span of the implant. The reason for this is not clear yet, however, the more spiky morphology may promote faster matrix production and subsequent mineralisation. It is generally thought that a roughened surface is preferential to strong bone bonding at the tissue-implant interface[14, 31]. In addition, it has been reported that spark plasma sintering[32] also affect the cellular responses; and the presence of soluble calcium phosphate of the MHA will also impact cell proliferation and differentiation[33]. Therefore, further studies are needed to separate the sintering effect and impurity phase effect by fabricating both NHA and MHA structure by SPS process.

5. CONCLUSION

In vitro cell culture showed that HOBs attached the surface of NHA with a lot of filopodia after 90min incubate, cell density on NHA surface is significantly higher than that of MHA surface and Glass control after 90 min incubation, whereas the average area of a spread cell after 4 h incubation is significantly lower on NHA surface than that on MHA surface and glass control, which might be attributed to the higher roughness, surface area and wettability of NHA surface, since such difference will affect the protein adsorption on the surface. After 14 days culture, NHA showed a significant enhancement ($p<0.05$) in matrix mineralisation compared to MHA. In general, NHA will not only promote the first stage of cell attachment, adhesion and spreading, but also improve the long-term cell proliferation and differentiation.

Acknowledgements
This work was partially supported by the Swedish research Council through grant 621-2005-6290.

REFERENCE

[1]Deligianni, D. D., Katsala, N. D., Koutsoukos, P. G., and Missirlis, Y. F., Effect of surface roughness of hydroxyapatite on human bone marrow cell adhesion, proliferation, differentiate and detachment strength. *Biomaterials*, **22**, 87-96(2001).

[2]Anselme, K., Osteoblast adhesion on biomaterials. *Biomaterials*, **21**, 667-81(2000).

[3]Anderson, J. M., Gristina, A. G., Hanson, S. R., Harker, L. A., Johnson, R. J., Merritt, K., Naylor, P. T., and Schoen, F. J., *Chapter 4 Host reactions to biomaterials and their evaluation*. 1996, Academic Press, San Diego.

[4]Thomas, C. H., McFarland, C. D., Jenkins, M. L., Rezania, A., and Healy, K. E., The role of vitronectin in the attachment and spatial distribution of bone derived cells on materials with patterned surface chemistry. *J. Biomed. Mat. Res.*, **37**, 81-93(1997).

[5]Webster, T. J., Nanophase ceramics: the future orthopedic and dental implant materials. *Advances in chemical engineering*, **27**, 125-66(2000).

[6]Barrerea, F., Snel, M. M. E., van Blitterswijk, C. A., de Groot, K., and Layrolle, P., Nano-scale study of the nucleation and growth of calcium phosphate coating on titanium implants. *Biomaterials*, **25**, 2901-10(2004).

[7]Klabunde, K. J., Stark, J., Koper, O., Mohs, C., Park, D., Decker, S., Jiang, Y., Lagadic, I., and Zhang, D., Nanocrystals as stoichiometric reagents with unique surface chemistry. *J. Phys. Chem.*, **100***(30)*, 12142-53(1996).

[8]Damien, C. J., and Parsons, J. R., Bone graft and bone braft substitutes: A review of current technology and applications. *Jouranl of applied biomaterials*, **2**, 187-208(1991).

[9]Vallet-Regi, M., and Gonzalez-Calbet, J. M., Calcium phosphate as substitution of bone tissue. *Progress in solid state chemistry*, **32**, 1-31(2004).

[10]Webster, T. J., Ergun, C., Doremus, R. H., Siegel, R. W., and Bizios, R., Enhanced functions of osteoblasts on nanophase ceramics. *Biomaterials*, **21***(17)*, 1803-10(2000).

[11]Webster, T. J., Siegel, R. W., and Bizios, R., Osteoblast adhesion on nanophase ceramics. *Biomaterials*, **20**, 1221-27(1999).

[12]Panda, R. N., Hsieh, M. F., Chung, R. J., and Chin, T. S., FTIR, XRD, SEM and solid state NMR investigations of carbonate-containing hydroxyapatite nano-particles synthesized by hydroxide-gel technique. *Journal of Physics and chemistry of Solids*, **64**, 193-99(2003).

[13]Wilson, K., and Goulding, K. H., *A biologist's guide to principles and techniques of practical biochemistry*. 1986, Edward Arnold.

[14]Gough, J. E., Notingher, I., and Hench, L. L., Osteoblast attachment and mineralized nodule formation on rough and smooth 45S5 bioactive glass monoliths. *J Biomed Mater Res*, **68A***(4)*, 640-50(2004).

[15]Mosmann, T., Rapid colorimetric assay for cellular growth and survival: application to proliferation and cytotoxicity assays. *J Immunol Methods*, **65***(1-2)*, 55-63(1983).

[16]Gloeckner, H., Jonuleit, T., and Lemke, H.-D., Monitoring of cell viability and cell growth in a hollow-fiber bioreactor by use of the dye Alamar Blue™. *Journal of Immunological Methods*, **252**, 131-38(2001).

[17]Nakayama, G. R., Caton, M. C., Nova, M. P., and Parandoosh, Z., Assessment of the Alamar blue assay for cellular growth and viability in vitro. *Journal of Immunological Methods*, **204**, 205-08(1997).

[18]Stanford, C. M., Jacobson, P. A., Eanes, E. D., and Lembke, L. A., Rapidly forming apatite mineral in an osteoblastic cell line. *The journal of biological chemistry*, **270***(16)*, 9420-28(1995).

[19]Gough, J. E., Jones, J. R., and Hench, L. L., Nodule formation and mineralisation of human primary osteoblasts cultured on a porous bioactive glass scaffold. *Biomaterials*, **25**, 2039-46(2004).

[20]Narayan, R. J., Kumta, P. N., Sfeir, C., Lee, D.-H., Olton, D., and Choi, D., Nanostructured ceramics in medical devices: Applications and prospects. *JOM*, **56***(10)*, 38-43(2004).

[21]Gough, J. E., Schotchford, C. A., and Downes, S., Cytotoxicity of glutaraldehyde crosslinked collagen/poly (vinyl alcohol) films is by the mechanism of apoptosis. *J. Biomed Mater Res*, **61**, 121-30(2002).

[22]Hammarback, J. A., McCarthy, J. B., Palm, S. L., Furcht, L. T., and Letourneau, P. C., Growth cone guidance by substrate bound laminin pathways is correlated to neuron-to-pathway adhesivity. *Dev Biol*, **126**, 29-39(1988).

[23]Dalby, M. J., Di-Silvio, L., Gurav, N., Annaz, B., Kayser, M. W., and Bonfield, W., Optimizing HAPEX™ topography influences osteoblast response. *Tissue Eng*, **8**, 453-67(2002).

[24]Schakenraad, J. M., Cells: Their surfaces and interactions with materials, in *Biomaterials Science: an introduction to materials in medicine*. 1996, Ratner, B. D., Hoffman, A. S., Schoen, F. J., and lemons, J. E., Eds., Academic Press, London.

[25]Dickenson, R. B., and Tranquillo, R. T., A stochastic model for adhesion-mediated cell random motility and haptotaxis. *J. Math Biol*, **31**, 563-600(1993).

[26]Lauffenburger, D. A., Models for receptor-mediated cell phenomena: adhesion and migration. *Ann. Rev. Biophys Chem*, **20**, 387-414(1991).

[27]Olbrich, K. C., Anderson, T. T., Blumenstock, F. A., and Bizios, R., Surface modified with covalently-immobilized adhesive peptides affect fibroblast population motility. *Biomaterials*, **17**, 759-64(1996).

[28]Healy, K. E., Rezania, A., and Stile, R. A., Designing biomaterials to direct biological responses. *Ann. NY Acad Sci*, **875**, 24-35(1999).

[29]Bearingger, J. P., Castner, D. G., and Healy, K. E., Biomolecular modification of p(Aam-co-EG/AA) IPNs supports osteoblast adhesion and phenotypic expression. *J. Biomater Sci Polym Ed*, **7**, 652-92(1998).

[30]Dee, K. C., Anderson, T. T., and Bizios, R., Design and function of novel osteoblast-adhesive peptites for chemical modification of biomaterials. *J Biomed Mater Res*, **40**, 371-77(1998).

[31]Gray, C., Boyde, A., and Jones, S. J., Topographically induced bone formation in vitro: Implication for bone implants and bone grafts. *Bone*, **18**, 115-23(1996).

[32]Nakahira, A., Tamai, M., Aritani, H., Nakamura, S., and Yamashita, K., Biocompatibility of dense hydroxyapatite prepared using an SPS process. *J. Biomed Mater Res*, **60***(4)*, 550-57(2002).

[33]Ogata, K., Imazato, S., Ehara, A., Ebisu, S., Kinomoto, Y., Nakano, T., and Umakoshi, Y., Comparison of osteoblast response to hydroxyapatite and hydroxyapatite/soluble calcium phosphate composites. *J. Biomed Mater Res*, **72A**, 127-35(2005).

OSTEOBLAST RESPONSE TO ZINC-DOPED SINTERED β-TRICALCIUM PHOSPHATE

Sahil Jalota, Sarit B. Bhaduri, and A. Cuneyt Tas
School of Materials Science and Engineering
Clemson University
Clemson, SC 29634

ABSTRACT

Sintered β-tricalcium phosphate (β-TCP, β-$Ca_3(PO_4)_2$) and Zn-doped (600, 4100, and 10100 ppm) β-TCP samples were prepared by using an aqueous chemical synthesis technique, followed by the calcination of pressed powders at 1000°C in air. Precursor powders of the synthesis process were Ca-deficient nanoapatites (i.e., Ca/P molar ratio varying from 1.49 to 1.51) with rod-like but agglomerated particles of 50 nm length and 20 nm thickness. In vitro culture tests performed by mouse osteoblast-like cells showed that β-TCP doped with 4100 ppm Zn had the highest cell viability and alkaline phosphatase (ALP) activity values over a range of 0 to 1 wt% Zn. The sample surface roughness, measured by non-contact profilometry, was also found to have an effect on the Live/Dead cell counts, and the highest cell viability recorded in this study corresponded to the surfaces with the least roughness.

INTRODUCTION

The most commonly used synthetic bone implant materials are Ca-hydroxyapatite (HA, $Ca_{10}(PO_4)_6(OH)_2$) and β-tricalcium phosphate (β-TCP, β-$Ca_3(PO_4)_2$).[1] These materials possess exceptionally good tissue compatibility and bond directly to bone without an intermediary layer of fibrous tissue.[2] Calcium phosphate (CaP)-based synthetic implants provide, *in vivo*, calcium and phosphate ions to the implant-host bone interface as soon as their resorption starts.[2] The inorganic part of bone is made up of a defective and rather complex substance (also doped with several mono- or divalent cations (Na, K, Mg, Zn, Fe, etc.) as well as with carbonate ions) with a generic formula of $Ca_{8.3}(PO_4)_{4.3}(HPO_4, CO_3)_{1.7}(OH, CO_3)_{0.3}$.[3] Divalent cations, which partially substitute the calcium and phosphate sites in these implant structures, seem to play an important role in the competition between HA and β-TCP.[4] However, β- TCP has an advantage over HA in the sense that β-TCP dissolves and resorbs faster than HA. It was shown that the dissolution rate of β-TCP (i.e., 1.26×10^{-4} mol/m^{-2} min^{-1}) in an aqueous solution at a pH of about 6 was about 89 times greater than that of carbonated apatite (1.42×10^{-6} mol/m^{-2} min^{-1}).[5] As an implant the higher dissolution rate of β-TCP may result in premature loss of mechanical strength. It was shown that when the TCP structure was stabilized, the dissolution rate will decrease,[6, 7] providing better mechanical properties. This stability in structure can be achieved by substituting the larger Ca^{2+} (0.099 nm) ions with smaller divalent cations, such as Zn^{2+} (0.074 nm)[6, 8] or Mg^{2+} (0.072 nm) ions.[7] It was reported that the solubility activity product (K_{sp}) of single phase pure β-TCP is 2.51×10^{-30} [9, 10] and that of Zn-doped TCP decreases by 52-92% in increasing the Zn-level up to 630 ppm.[6] Stabilization of the β-TCP structure was quite evident from the fact that there was an increase in the transformation temperature of Zn-doped β-TCP to α-TCP.[11] As a result, Zn-doped β-TCP can be sintered at higher temperatures without its conversion to α-TCP.

Zinc is an important growth factor, as deficiency of zinc can adversely affect growth in many animal species and in humans.[12] The deficiency of zinc can also cause severe disorders as poor appetite, mental lethargy, delayed wound healing, growth retardation, delayed puberty in adolescents, and rough skin.[12] The supplementation of zinc can help reduce the susceptibility to diseases like diarrhea, pneumonia, respiratory infections, and poor immune system.[13] Zinc is essential for maintaining biologically good health in humans. In humans, zinc is present as a trace element in the bones, teeth, hair, skin, testes, liver, and muscles. Zinc also promotes synthesis of deoxyribonucleic acid (DNA) and ribonucleic acid (RNA). Zinc has a stimulatory effect on bone formation and mineralization, both in vivo and in vitro.[14-16] The presence of zinc is known to increase the protein synthesis,[17-19] to activate the aminoacyl-tRNA synthase,[19, 20] to enhance albumin synthesis,[18] and to increase ALP activity.[15] It should be noted here that a protein called osterix (i.e., zinc finger-containing transcription factor) is needed for osteoblast differentiation.[21] While it was shown that the presence of zinc led to an increase in osteoblast activity, it was also shown that zinc's deficiency led to bone growth retardation,[22,23] postmenopausal osteoporosis,[24] and programmed cell apoptosis (in mice).[25] However, there comes a limitation in the amount of zinc which can increase the activity of osteoblast cells, as high concentrations of zinc may also have a cytotoxic effect on cells.[26] Therefore, in this study we placed the significance on finding the appropriate zinc concentration needed for enhanced osteoblast cell proliferation, response and growth.

The synthesis of zinc containing calcium phosphates has been initialized by Bigi *et al.*[4] and Fuierer *et al.* [27] Bigi *et al.* synthesized Zn-doped β-TCP by physically mixing the in-house synthesized single phase β-TCP with α-$Zn_3(PO_4)_2$, followed by calcining the mixture at 1000°C for 15 hours.[8] In the aforementioned studies no ICP data were reported which would have been of great help in determining any vaporization of Zn occurring while heating the samples at such high temperatures for prolonged times. Another important contribution to the synthesis of Zn-doped β-TCP was made by LeGeros *et al.*,[28,29] where the precursor powders were formed by wet chemical methods. However, until now the research group of Ito *et al.*[6, 30-42] have been the most important contributor to the synthesis of Zn-doped β-TCP (and other calcium phosphates), as well as their *in vitro* and *in vivo* evaluation. Ito *et al.* have shown the positive stimulatory effects of Zn on *in vivo* bone formation. The most preferred route of the Ito group in synthesizing the Zn-doped β-TCP was a two-step procedure.[31] Briefly, they first prepared a 10 mol% Zn-doped β-TCP and then mixed it in an alumina mortar with appropriate amounts of commercially available pure β-TCP to obtain the desired amount of Zn-doping. 10 mol% Zn-doped β-TCP was synthesized by mixing a suspension of calcium hydroxide (synthesized in-house by forming calcium oxide from calcium carbonate by heating the latter at 1000°C for 3 h and then dissolving the former in water) with zinc nitrate hexahydrate and a phosphoric acid solutions.[31] The precipitates were then filtered and calcined at 850°C for 1 h.[31] The authors also reported the presence of a secondary phase of $CaZn_2(PO_4)_2$ while attempting to synthesize TCP containing zinc more than 12 mol%. After mixing 10 mol% Zn-doped β-TCP with pure β-TCP, various samples were reported to be synthesized, such as containing 0.28, 2.56, 5.0, 7.47, and 10.5% Zn.[31]

We realized that the lower end of Zn-doping range was not fully explored. Therefore, in this study we tried to explore the range of Zn-doping from pure β-TCP to 1.0 wt% Zn-doped β-TCP. We synthesized the powders by using an aqueous chemical route which used $Ca(NO_3)_2 \cdot 4H_2O$, $NH_4H_2PO_4$, and $Zn(NO_3)_2 \cdot 6H_2O$ as the starting chemicals. Precursor powders were pressed into pellets and then calcined at 1000°C for 6 h and characterized by using analytical techniques. The pellets were then tested for their cell viability and ALP activity using rat osteoblast cells.

EXPERIMENTAL PROCEDURE

In this work, we studied the effect of varying concentrations of Zn in β-TCP to determine the osteoblast response as a function of Zn concentration. The samples prepared were pure β-TCP, and 600 ppm, 4100 ppm, and 10100 ppm Zn-doped β-TCP. For synthesizing these Zn-containing calcium phosphates, $Ca(NO_3)_2{\cdot}4H_2O$, $NH_4H_2PO_4$, and $Zn(NO_3)_2{\cdot}6H_2O$ (Reagent-grade, Fisher Chemicals, Fairlawn, NJ) were used. The (Ca+Zn)/P molar ratio was maintained at 1.503 for all the samples and two aqueous solutions were prepared; one containing a phosphate salt dissolved in deionized water and the second containing Ca and Zn salts dissolved in water. For synthesizing 0/ 600/ 4100/ 10100 ppm Zn-doped β-TCP, the first solution contained 0.1951 moles $NH_4H_2PO_4$, a constant amount for all concentrations. The second solution was prepared by dissolving 0.2932/ 0.2924/ 0.2910/ 0.2877 moles of $Ca(NO_3)_2{\cdot}4H_2O$ and 0 / 0.0008/ 0.0022/ 0.0055 moles of $Zn(NO_3)_2{\cdot}6H_2O$, respectively, in 600 mL water. The latter solution containing Ca^{2+} and Zn^{2+} ions was rapidly added to the phosphate solution. Within few minutes after addition, the solution became turbid and pH was recorded as 4±0.1. To make this solution clear, few drops of conc. HNO_3 (15.69 M, Fisher) were added and the pH of this solution dropped to 3±0.1. This clear solution was then stirred at 37±1°C for 2 h followed by rapid addition of 50 mL NH_4OH (29% NH_3, Merck) causing instantaneous precipitation and a resultant opaque solution with a pH of 9.2±0.2 at 37°C. This suspension was stirred for 1 h and filtered with paper. The precipitates were washed with 4 L water, followed by drying at 90°C overnight. The dried powders were ground and then pressed into pellets using a 1.25 cm diameter steel die and a pressure of 4,500 kg. Thus formed pellets were calcined at 1000°C for 6 hours in air.

The precursor and calcined samples were characterized by using an X-ray diffractometer (XRD; XDS 2000, Scintag Corp., Sunnyvale, CA), operated at 40 kV and 30 mA with monochromated Cu K_α radiation. X-ray data were collected at 2θ values from 25° to 40° at a rate of 0.03°/minute. Fourier-transformed infrared spectroscopy (FTIR; Nicolet 550, Thermo-Nicolet, Woburn, MA) analysis was performed on the precursor and calcined samples. The size and shape of the precursor particles were evaluated by transmission electron microscope (TEM, H7600T, Hitachi Corp., Tokyo, Japan) at 120kV. Surface morphology of the sputter-coated (w/Pt) calcined pellets was evaluated with a scanning electron microscope (FE-SEM; S-4700, Hitachi Corp., Tokyo, Japan) which was used in the secondary electron (SE) mode with an acceleration voltage of 5 kV. Chemical analyses of powder (both precursor and calcined) samples were performed by ICP-AES (Model 61E, Thermo Electron, Madison, WI). For the ICP analyses, 50 mg portions of powder samples were dissolved in 5 mL of concentrated HNO_3 solution. Surface roughness analyses on calcined pellets were performed with a NT-2000 Non-contacting surface profilometer (Wyko, Tuscon AZ) with a 0.164 x 0.215 mm field of view and a magnification of 25X. In profilometry, R_a was the average roughness and R_t was the difference between of the highest peak and the lowest valley in the field of view. The bulk density measurements of calcined pellets were performed using He-pycnometer (AccuPyc 1330, Micromeritics, Norcross, GA). For each sample, the number of purges and runs was 5 and the averages were reported along with the standard deviation. Thermogravimetric analyses (TGA, Model 851e, Mettler-Toledo, Columbus, OH) were performed in an air atmosphere only on the starting chemicals of our powder synthesis route over the range of 30°–1000°C, with a scan rate of 5°C/min.

7F2 rat osteoblast cells (CRL-12557, American Type Culture Collection, Rockville, MD) were grown on 75 cm^2 culture flasks at 37°C and 5% CO_2 in alpha minimum essential medium (α-MEM) with 2 mM 1-glutamine and 1 mM sodium pyruvate, without ribonucleosides and deoxyribonucleosides, augmented by 10% FBS. The culture medium was changed every other day until the cells reached a confluence of 90-95%, as determined visually with an inverted microscope. The cells were passaged using trypsin (2.5 g/L)/ EDTA (25mM) solution (Sigma-Aldrich Corp., St. Louis, MO, USA). The obtained cells were then seeded at a concentration of 3500 cell/well on 0.14 cm^3 cylindrical samples for various assays. Cell viability and alkaline phosphatase activity were measured after 72 hours. For statistics, the sample size (n) was selected as 16 for all the *in vitro* cell culture tests.

The cell viability assessment was performed using Live/Dead® Viability/Cytotoxicity Kit (L-3224, Molecular Probes, Eugene, OR). The fluorescence values at 494/517 nm for live cells and 528/617 nm for dead cells were recorded. The alkaline phosphatase (ALP) activity was determined using the ALP concentration and the cell extracted protein concentration. The ALP concentration was calculated using Enzymatic Assay of Phosphatase Alkaline Kit (EC 3.1.3.1, Sigma-Aldrich Corp., St. Louis, MO, USA). A working reagent was prepared by first mixing 2.7 ml of Reagent A (1.0 M Diethanolamine Buffer with 0.50 mM Magnesium Chloride) with 0.30 ml of Reagent B (150 mM p-Nitrophenyl Phosphate Solution (pNPP)) and then mixing the mixture with 0.10 ml of cell-containing media. 100 μl of this solution was added to each well and thoroughly mixed and incubated at 37°C for 30 minutes. Following incubation, the absorbance was recorded at 405 nm with the spectrophotometer at room temperature. The standard curve was obtained by plotting the absorbance measured at 405 nm for certain concentration against the concentration in μg/ml. ALP concentration of each sample was then determined using this standard curve and is expressend as μg-pNP/ml. The cell extracted protein concentration was determined in a two-step procedure, first the protein was extracted using M-PER™ Mammalian Protein Extraction reagent and then this extracted protein was measured using BCA™ Protein Assay Kit. The cell samples were lysed by adding 200 μl of M-PER™ Reagent to each well plate and then shaking for 5 minutes. Lysate was collected and transferred to microcentrifuge tubes, followed by centrifuging at 4000g for 10 minutes to pellet the cell debris. Supernatant was transferred to clean tubes for analyzing the protein concentration. To measure the protein amount, a working reagent (WR) was prepared by mixing 50 parts of BCA™ Reagent A with 1 part of BCA™ Reagent B (50:1, Reagent A:B). 200 μl of the above mentioned WR was added to each well and thoroughly mixed. Following mixing, the well plate was covered and incubated at 37°C for 30 minutes. The absorbance at 562 nm was measured with the spectrophotometer at room temperature. A standard curve was prepared by plotting the average blank-corrected 562 nm measurement for each BSA standard versus its concentration in μg/ml. Cell extracted protein concentration was then determined by using this standard curve and is expressed as μg/ml. The ALP activity was then calculated as follows; ALP Activity = [(μg pNP)/139] / μg(cell extracted protein) = μmoles pNPP/ μg cell protein. Osteoblast attachment on calcined pellets was examined using SEM (FESEM; S-4700, Hitachi Corp., Tokyo, Japan). Prior to the SEM investigations, cells were fixed by using 3.5% glutaraldehyde. Osteoblasts were dehydrated through sequential washings in 50%, 70%, 95% ethanol solutions and 2 times in 100% ethanol. Samples were then critical point-dried according to the previously published techniques. Samples were sputter-coated with a thin layer of platinum prior to the electron microscope observations performed at 5 kV.

RESULTS AND DISCUSSION

The ICP-AES results (Table 1) showed the Ca, P and Zn levels achieved in the powders after calcinations. Henceforth, the samples will be referred to as Zn-0, Zn-600, Zn-4100, Zn-10100 for 0, 600, 4100, 10100 ppm levels of Zn in β-TCP. It must be noted that there was an increase in Zn levels in going from precursors to 1000°C-calcined samples (results not shown),

Table I ICP-AES results of the calcined samples at 1000°C for 6 h

Sample *	Ca (%)	P (%)	Ca/P	Zn (*ppm*)
Zn-0	41.21	20.71	1.523	0
Zn-600	42.85	22.14	1.496	600
Zn-4100	36.46	18.94	1.488	4100
Zn-10100	38.92	20.16	1.493	10100

* Samples containing ppm-level Zn in β-TCP

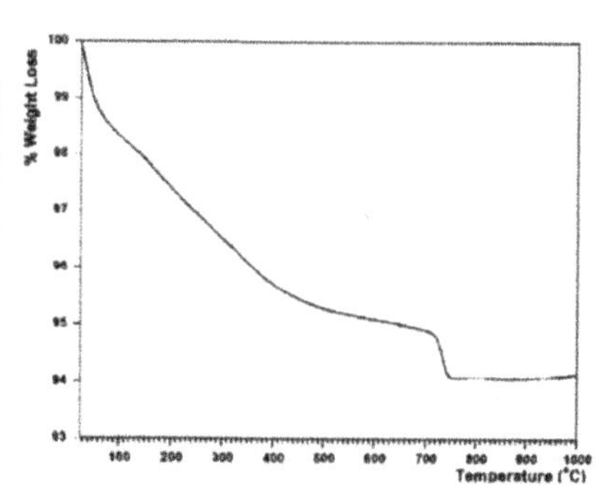

Fig. 1a

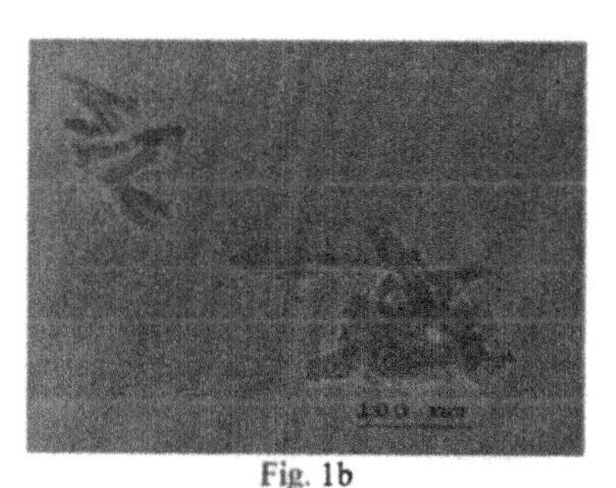

Fig. 1b

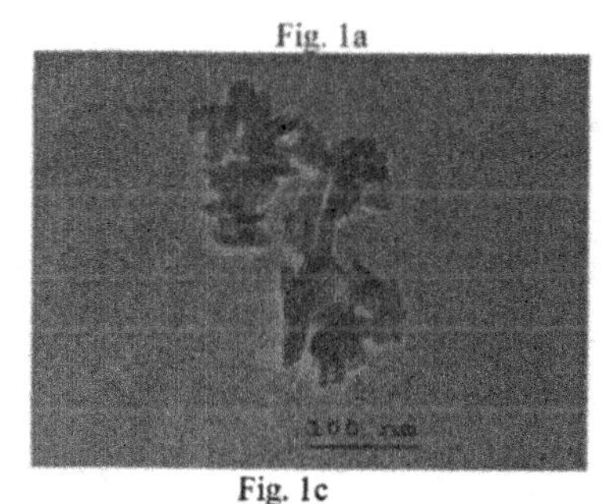

Fig. 1c

Fig. 1 (*a*) TGA trace, (*b*, *c*) TEM micrographs of the precursor powders

which can be attributed to the different forms (i.e., surface/bulk) of Zn present in both sets of samples. Zinc will be more tightly incorporated into the structure after calcination as compared to the surface-adsorbed Zn that might be present in the precursor powders. A total weight loss of around 6% was observed in going from room temperature (RT) to 1000°C (Fig. 1a). The weight loss occurred in two parts, first comprised of a gradual loss from RT to 700°C which meant a loss of some surface adsorbed water, remaining nitrate and/or ammonium ions which were not fully removed during washing. The conversion of hydrogen phosphate ions (HPO_4^{2-}) into $P_2O_7^{4-}$ occurs at the same time, with the evolution of water vapor. Carbonate ions that were present in the FTIR spectra of the precursor material and absent in the calcined samples, constituted the second part. These carbonate ions were removed from the system at around 720°C.

The TEM micrographs (of Figures 1b and 1c) of the precursor material showed that the material was nanocrystalline in nature and consisted of rod-shaped crystals of about 50 nm-length and 15 nm-thickness. Due to the high concentration of the powders in solution, these crystals tended to form agglomerates. These agglomerates were believed to be soft agglomerates as these were later on pressed to form a pellet. Although the sample preparation routes of this study differed significantly from that of Bouyer *et al.*,[13] the particle shape and size was quite similar.

XRD patterns of the pure and Zn-doped precursor powders and sintered samples were shown in Figure 2a. The nanorods or nanoneedles of the precursor powder yielded an XRD pattern similar to that of poorly-crystallized apatitic calcium phosphate. Sintered samples (both

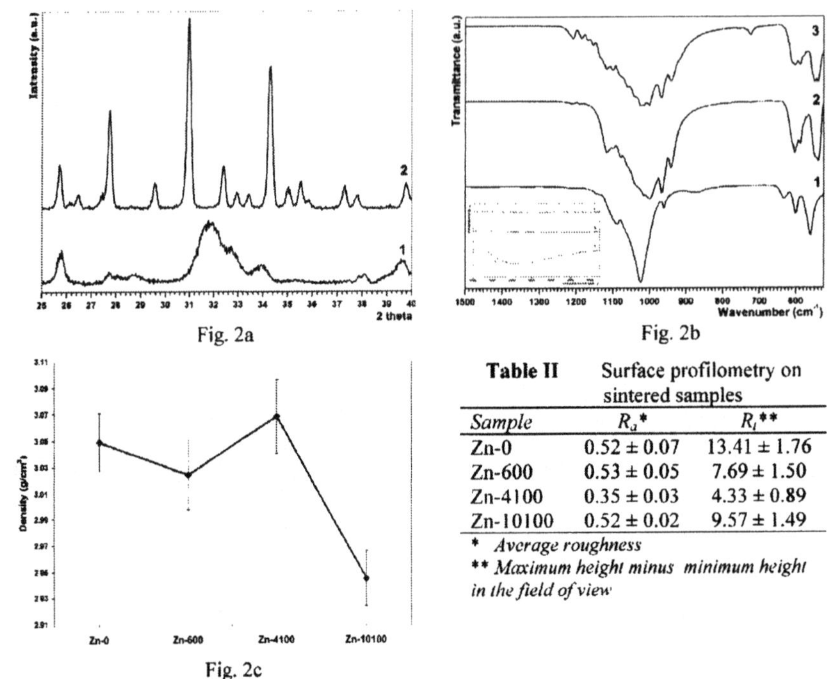

Table II Surface profilometry on sintered samples

Sample	R_a*	R_t**
Zn-0	0.52 ± 0.07	13.41 ± 1.76
Zn-600	0.53 ± 0.05	7.69 ± 1.50
Zn-4100	0.35 ± 0.03	4.33 ± 0.89
Zn-10100	0.52 ± 0.02	9.57 ± 1.49

* *Average roughness*
** *Maximum height minus minimum height in the field of view*

Fig. 2c

Fig. 2 (*a*) XRD trace, and (*b*) FTIR trace of the precursor powder (1), sintered sample (2), and Fluka TCP powder, (*c*) pycnometric results (bulk densities) of sintered pellets

pure and Zn-doped) resulted in the characteristic β-TCP XRD patterns. The FTIR spectra (Fig. 2b) confirmed the X-ray analyses, where a band at 3571 cm^{-1} was observed (see inset in Fig. 2b) substantiating the presence of apatitic phase in the precursors. The OH^- band in the sintered samples was absent, indicating the conversion to β-TCP. The characteristic IR spectrum for the commercial β-TCP powders (i.e., Fluka, Inc.) was added as trace 3 to Fig. 2b. The IR bands of trace 2 and trace 3 of this chart match closely. The precursors after pelletization and calcination resulted in highly dense pellets, the bulk density values reported in Figure 2c. The bulk density of β-TCP is 3.15 gm/cc, which confirms the attainment of more than 95% density in Zn-doped β-TCP. The maximum densification was observed in the sample of Zn-4100. Interestingly, this sample (Zn-4100) exhibited the lowest average surface roughness among all the samples. Average surface roughness and highest-to-lowest peak values are reported in Table 2, which showed the lowest values for Zn-4100, thus the sample with the smoothest surface. This effect was quite evident from the SEM micrographs of sintered pellets (shown in Figure 3), where a decrease in grain size was observed when Zn levels were either increased or decreased from 4100 ppm. The grain sizes were ranging from 350 nm to 2 μm for Zn-4100 and ranged from a 100 nm to < 1μm for the rest of Zn-dopant levels. This grain growth as a function of substitution of divalent ions into β-TCP was also observed by Yoshida *et al.*[44] They reported an increase in the grain size until 11.5 mol% Mg^{2+} ion substitution and observed a decrease in the grain size thereafter. This case resembles our study in which instead of Mg^{2+} we used Zn^{2+} and that the maximum grain growth was observed at 4100 ppm Zn-doped β-TCP samples, with a slight decrease in grain sizes thereafter.

In β-TCP, Ca(4) and Ca(5) sites are the most suitable sites for smaller cation substitution.[45] The site Ca(4) was confirmed to be very different from the other four calcium sites with a lower occupancy factor, higher isotropic thermal parameter, a small bond valence sum (BVS), small co-ordination number (CN) and longer Ca-O distances.[46] Ca(5) site is co-ordinated with six oxygen atoms and has the highest BVS of 2.7 as compared to the others.[46] The other cation sites do not offer a suitable geometry for small cations. Theoretical evidence has proved that when Zn^{2+} ion substitution took place, a decrease in bond lengths, an increase in chemical stability and a deformation of the crystal structure of hydroxyapatite occurred.[47] A similar effect was also observed when Mg^{2+} ions were substituted in HA.[47] In Mg-doped β-TCP, it was observed that the Mg(4)•••O(9) bond was shortened in comparison to the Ca(4)•••O(9) bond in pure β-TCP.[45, 48] The O•••Mg(5)•••O bonds approached towards 90° with increasing Mg content confirmed the trend towards the more ideal octahedral configuration.[45, 48] Moreover, due to the similarity in the lattice constants of Zn and Mg, a similar preferential distribution in different cationic sites is expected from both the theoretical and experimental studies.[8] Therefore, zinc substitution in β-TCP will lead to shortening of the bond lengths and an increase in chemical stability.

Mouse osteoblast cells (7F2) cultured on calcined pellets exhibited differences in terms of the number of attached cells and alkaline phosphatase activity, as presented in Figures 4a and 4b. The initial number of cells seeded onto the calcined pellets was 3500 cells/pellet and after 3 days the number of live cells on Zn-0 sample increased to about 15000 cells. All the other Zn-doped pellets had greater number of live cells, with the highest being at 4100 ppm Zn-dopant level. Interestingly, Zn had a cytotoxic effect and it was seen that the number of cell death was

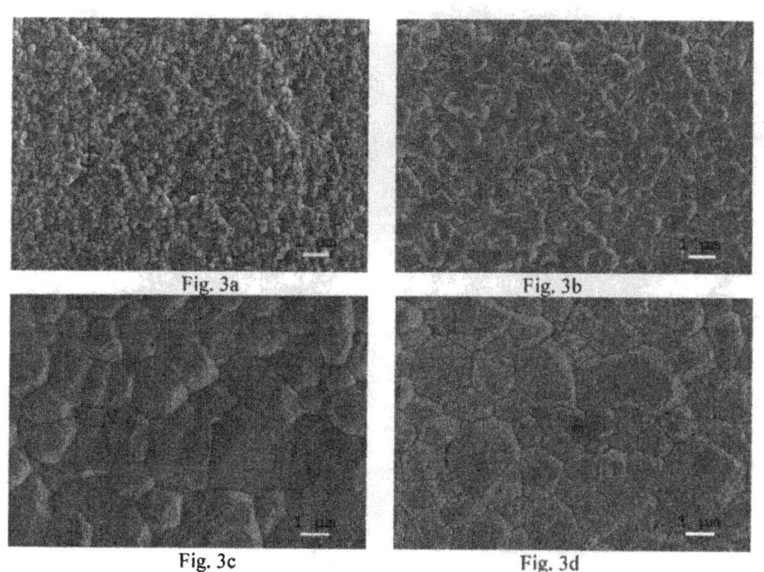

Fig. 3 SEM micrographs of (*a*) pure β-TCP, (*b*) Zn-600, (*c*) Zn-4100, and (*d*) Zn-10100 pellet

maximum on the pellets which contained the highest amount of Zn-level in this study, that was, Zn-10100. The % dead cells were calculated for all the samples, and a descending trend was observed with Zn-10100 sample showing the highest % dead cells at around 21% and the least was observed for pure-TCP sample (4%). Therefore, zinc stimulated the growth of osteoblasts and we observed that the osteoblasts multiplied by ten-fold within 3 days in Zn-doped samples, whereas the multiplication rate was only 5-fold in Zn-0 (pure -TCP). The osteoblast cells require the presence of certain ions and proteins (i.e., nutrients) in the media for growth and propagation.[49] We may speculate that at the very beginning of the cell culture tests, the amount

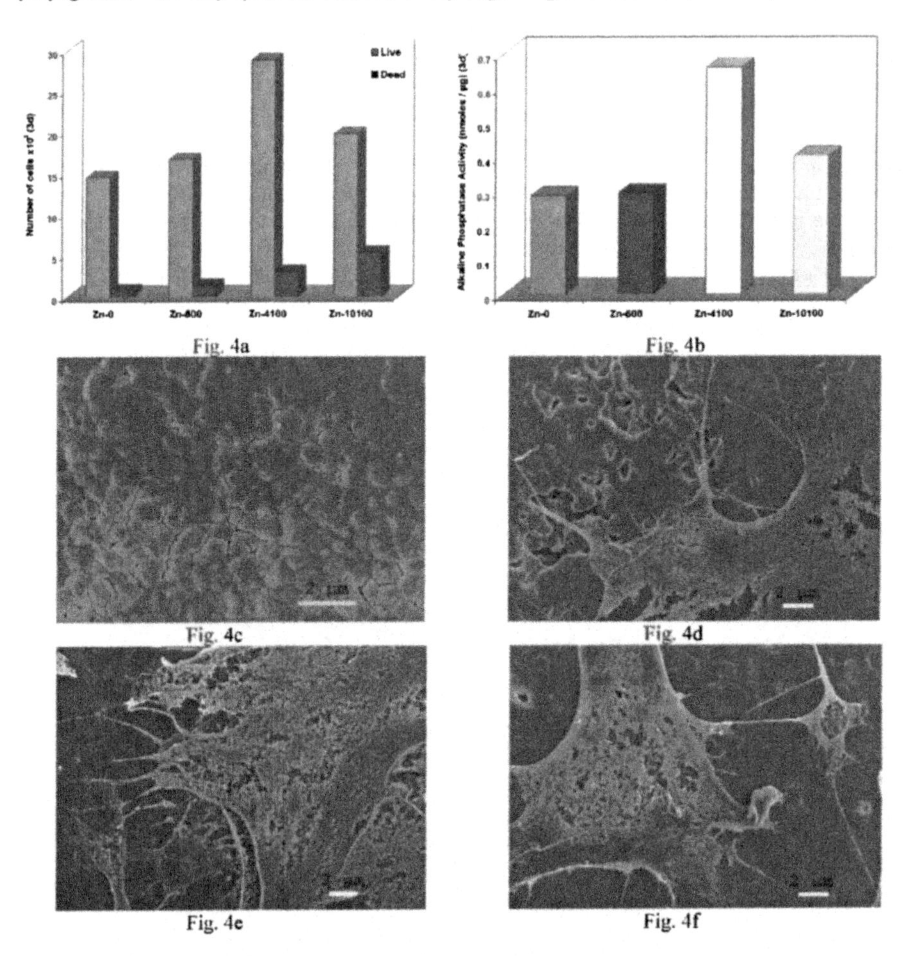

Fig. 4a

Fig. 4b

Fig. 4c

Fig. 4d

Fig. 4e

Fig. 4f

Fig. 4 (*a*) Cell viability (*live/dead*), (*b*) ALP activity for Zn-0, Zn-600, Zn-4100 and Zn-10100; SEM micrographs of osteoblasts on the surface of (*c*) pure β-TCP (*d*) Zn-600, (*e*) Zn-4100, and (*f*) Zn-10100 pellet

of ions in the media was sufficient which helped the cells to multiply. Since the media was not replenished during the 3- day culture tests, cell death was observed due to a decreased concentration of the essential ions and proteins in the media. The alkaline phosphatase activity was also found to yield the highest values for Zn-level of 4100 ppm in the sintered pellets. A decrease in both the ALP activity and live cells was observed with an increase or decrease in Zn-levels from 4100 ppm. Osteoblast attachment and proliferation on the surfaces of sintered pellets was monitored by FESEM, and given in Figures 4c through 4f. Osteoblasts were attached to the surfaces of all the samples tested here, however, the osteoblast proliferation showed a significant behavior. Cells were able to easily differentiate between the chemical compositions of the sintered pellets. In Zn-0 pellet (pure β-TCP), the osteoblast cell was observed to be closely associated with the pellet, as they formed a translucent layer through which the grains were still visible (Fig. 4c). In Zn-doped pellets, osteoblasts appeared to be more elongated with numerous filopodia extending onto the pellet surfaces. The cells were also slightly raised above the surface in these samples as compared to the flattened morphology observed for the pure β-TCP pellets. Briefly, the extent of cell spreading was the highest in Zn-0 (pure β-TCP) pellets.

Factors like surface roughness and composition of a surface influence the material-cell interactions.[51, 52] Many researchers claimed that on smooth surfaces, better cell adhesion and spreading should be observed.[52, 53] However, there are discrepancies in literature and some studies demonstrated that greatest osteoblast attachment and proliferation was observed on rougher surfaces with more irregular topographies.[54, 55] An SEM study performed by Baxter *et al.*[50] to evaluate the material-cell interaction concluded that the spreading of the osteoblasts was highest on Thermanox plastic as compared to that of HA substrates. Interestingly, in another study, it was shown that proliferation numbers and alkaline phosphatase (ALP) activities of the osteoblasts on the composite calcium phosphate coatings were improved by approximately 30-40% when compared to Thermanox control, and were comparable to the pure HA coating.[56] These suggest that the *in vitro* observation of a high cell spreading would not necessarily mean better *in vivo* bone integration for that material.[50] In our study, we observed a mixed response in terms of osteoblast spreading and activity. The best spreading was observed on pure β-TCP (roughest surface in this study) while the best osteoblast activity was observed on a Zn-doped pellet with the smoothest surface, that is, at a Zn-dopant level of 4100 ppm. Higher levels of Zn-doping apparently led to the onset of cytotoxic effects (Fig. 4a), but still the ALP activity was higher (Fig. 4b) and the surfaces of those pellets were coarser than those of pure β-TCP.

In consequence, these results suggested that material composition, cell viability and ALP activity were the factors which are more important than the surface roughness and extent of cell spreading alone, in evaluating the interaction of osteoblast cells with the biomaterial under question. Evidence presented here largely added to the previous work performed by Ito *et al.*,[30-42] and partially satisfied our quest for the determination of the appropriate Zn dopant level in β-TCP to improve the osteoblast response.

CONCLUSIONS

A wet chemical method was developed and successfully used for the synthesis of β-TCP and Zn-doped β-TCP (0, 600, 4100 and 10100 ppm Zn), after carefully and extensively checking the Zn-dopant levels achieved in all the resultant samples by using an accurate chemical analysis technique like ICP, for the first time. Zn-doping, as well as the use of precursor powders consisting of nanorods or nanoneedles, significantly improved the densification of β-TCP at a relatively low temperature of 1000°C. Zn-doping caused a slight grain growth, in comparison to pure β-TCP, up to a certain dopant level, which then decreased or leveled off with further increase in the dopant level. The highest number of live rat osteoblast cells was observed for the 4100 ppm Zn-doped β-TCP pellet. This proved that even the ppm level presence of Zn (solely originating from the sintered pellet samples themselves) in a cell culture medium has an undeniable effect in stimulating the multiplication and proliferation of osteoblasts. The highest alkaline phosphatase activity was again encountered for the same Zn-4100 sample. Doping of sintered β-TCP porous blocks, sponges or granules, which are already in clinical use as bone substitutes, with Zn over the range of 3000 to 4000 ppm may cause a significant increase in the osteoblast response and proliferation on such samples. This has been the most critical and technologically important conclusion of this study.

ACKNOWLEDGEMENTS

The work is partially supported by NSF 0522057.

REFERENCES

[1] J.C. Heughebaert, and G. Bonel, "Composition, stuctures and properties of calcium phosphates of biological interest," In: Christel P. Meunier A, Lee AJC (eds): *Biological and Biomechanical Performance of Biomaterials* (Elsevier Science Publishers), 9-14 (1986).

[2] C.E. Rawlings, J.A. Persing, and J.M. Tew, "Modern bone substitutes with emphasis on calcium-phosphate ceramics and osteoinductors," *Neurosurgery*, **33**, 935-8 (1993).

[3] A. C. Tas, "Participation of calcium phosphate bone substitutes in the bone remodeling process: Influence of materials chemistry and porosity," *Key Eng. Mat.*, **264-268**, 1969-72 (2004).

[4] A. Bigi, E. Foresti, M. Gandolfi, and N. Roveri, "Inhibiting effect of zinc on hydroxylapatite crystallization," *J. Inorg. Biochem.*, **58**, 49-58 (1995).

[5] R. Tang, M. Hass, W. Wu, S. Gulde, and G. H. Nancollas, "Constant Composition of Mixed Phases II Selective Dissolution of Calcium Phosphates," *J. Colloid. Interf. Sci.*, **260**, 379-84 (2003).

[6] A. Ito, H. Kawamura, S. Miyakawa, P. Layrolle, N. Kanzaki, G. Treboux. K. Onuma, and S. Tsutsumi, "Resorbability and solubility of zinc-containing tricalcium phosphate," *J. Biomed. Mater. Res.*, **60**, 224-31 (2002).

[7] I. Manjubala, and T.S.S. Kumar, "Preparation of biphasic calcium phosphate doped with magnesium fluoride for osteoporotic applications," *J. Mater. Sci. Lett.*, **20**, 1225-7 (2001).

[8] A. Bigi, E. Foresti, M. Gandolfi, M. Gazzano, and N. Roveri, "Isomorphous substitutions in beta-tricalcium phosphate: The different effects of zinc and strontium," *J. Inorg. Biochem.*, **66**, 259-65 (1997).

[9] R. Tang, W. Wu, M. Haas, and G.H. Nancollas, "Kinetics of Dissolution of b-Tricalcium Phosphate," *Langmuir*, **17**, 3480-5 (2001).

[10] W. Wu, R. Tang, M. Haas, and G.H. Nancollas, "Constant Composition Dissolution Kinetics of Mixed Phases I. Synthetic Calcium Phosphates," *J. Coll. Interf. Sci.*, **244**, 347-52 (2001).

[11] E.R. Kreidler, and F.A. Hummel, "Phase equilibria in the system $Ca_3(PO_4)_2$ - $Zn_3(PO_4)_2$," *Inorg. Chem.*, **6**, 524-8 (1967).

[12] A.S. Prasad, "Clinical, endocrinological and biochemical effects of zinc-deficiency," *Clin. Endocrinol. Meta.*, **14**, 567-89 (1985).

[13] M. Hambidge, "Human zinc deficiency," *J. Nutr.*, **130**, 1344S-49S (2000).

[14] M. Yamaguchi, and R. Yamaguchi, "Action of zinc on bone metabolism in rats - increases in alkaline-phosphatase activity and DNA content," *Biochem. Pharmacol.*, **35**, 773–7 (1986).
[15] S.L. Hall, H.P. Dimai, and J.R. Farley, "Effects of zinc on human skeletal alkaline phosphatase activity in vitro," *Calcif. Tiss. Int.*, **64**, 163-72 (1999).
[16] D. Hatakeyama, O. Kozawa, T. Otsuka, T. Shibata, and T. Uematsu, "Zinc suppresses IL-6 synthesis by prostaglandin F-2 alpha in osteoblasts: Inhibition of phospholipase C and phospholipase D," *J. Cell. Biochem.*, **85**, 621-8 (2002).
[17] S.K. Lim, Y.J. Won, H.C. Lee, K. Huh, and Y.S. Park, "A PCR analysis of ER alpha and ER beta mRNA abundance in rats and the effect of ovariectomy," *J. Bone Miner. Res.*, **14**, 1189-96 (1999).
[18] E. Sugimoto and M. Yamaguchi, "Stimulatory effect of daidzein in osteoblastic MC3T3-E1 cells," *Biochem. Pharmacol.*, **59**, 471-5 (2000).
[19] K. Ishida, N. Sawada, and M. Yamaguchi, "Expression of albumin in bone tissues and osteoblastic cells: Involvement of hormonal regulation," *Int. J. Mol. Sci.*, **14**, 891-5 (2004).
[20] M. Yamaguchi, S. Kishi, and M. Hashizume, "Effect of zinc-chelating dipeptides on osteoblastic MC3T3-E1 cells - activation of aminoacyl-transfer-rna synthetase," *Peptides*, **15**, 1367-71 (1994).
[21] K. Nakashima, X. Zhou, G. Kunkel, Z. Zhang, J. M. Deng, R. R. Behringer, and B. de Crombrugghe, "The novel zinc finger-containing transcription factor Osterix is required for osteoblast differentiation and bone formation," *Cell*, **108**, 17-29 (2002).
[22] L.S. Hurley, J. Gowan, and G. Milhaund , "Calcium metabolism in. manganese deficient and zinc deficient rats," *P. Soc. Exp. Biol. Med.*, **130**, 856-60 (1969).
[23] G. Oner, B Bhaumick, and R.M. Bala, "Effect of zinc deficiency on serum somatomedin levels and skeletal growth in young rats," *Endocrinology*, **114**, 1860-3 (1984).
[24] M. Hertzberg, J. Foldes, R. Steinberg, and J. Menczel, "Zinc excretion in osteoporotic women.," *J. Bone Miner. Res.*, **5**, 251-7 (1990).
[25] P.J. Fraker, "Roles for cell death in zinc deficiency," *J. Nutr.*, **135**, 359-62 (2005).
[26] R.J.P. Williams, "Zinc: what is its role in biology?," *Endeavour*, **8**, 65-70 (1984).
[27] T.A. Fuierer, M. LoRe, S.A. Puckett, and G.H. Nancollas, "A mineralization adsorption and mobility study of hydroxyapatite surfaces in the presence of zinc and magnesium ions," *Langmuir*, **10**, 4721-5 (1994).
[28] R.Z. LeGeros, *Caries Res.* **31**, 434 (1997).
[29] R.Z. LeGeros, C.B. Bleiwas, M. Retino, R. Rohanizadeh, and J.P. LeGeros, "Zinc effect on the in vitro formation of calcium phosphates: Relevance to clinical inhibition of calculus formation," *Am. J. Dent.*, **12**, 65-71 (1999).
[30] N. Kanzaki, K. Onuma, G. Treboux, S. Tsutsumi, and A. Ito, "Inhibitory effect of magnesium and zinc on crystallization kinetics of hydroxyapatite (0001) face," *J. Phys. Chem. B.*, **104**, 4189-94 (2000).
[31] A. Ito, K. Ojima, H. Naito, N. Ichinose, and T. Tatseishi, "Preparation, solubility, and cytocompatibility of zinc-releasing calcium phosphate ceramics," *J. Biomed. Mater. Res.*, **50**, 178-83 (2000).
[32] H. Kawamura, A. Ito, S. Miyakawa, P. Layrolle, K. Ojima, N. Ichinose, and T. Tateishi, "Stimulatory effect of zinc-releasing calcium phosphate implant on bone formation in rabbit femora," *J. Biomed. Mater. Res.*, **50**, 184-90 (2000).

[33] M. Otsuka, S. Marunaka, Y. Matsuda, A. Ito, P. Layrolle, H. Naito, and N. Ichinose, "Calcium level-responsive in-vitro zinc release from zinc containing tricalcium phosphate (ZnTCP)," *J. Biomed. Mater. Res.*, **52**, 819-24 (2000).
[34] N. Kanzaki, K. Onuma, G. Treboux, S. Tsutsumi, and A. Ito, "Effect of impurity on two-dimensional nucleation kinetics: Case studies of magnesium and zinc on hydroxyapatite (0001) face," *J. Phys. Chem. B.*, **105**, 1991-4 (2001).
[35] A. Ito, H. Kawamura, M. Otsuka, M. Ikeuchi, H. Ohgushi, K. Ishikawa, K. Onuma, N. Kanzaki, Y. Sogo, and N. Ichinose, "Zinc-releasing calcium phosphate for stimulating bone formation ," *Mat. Sci. Eng. C-Bio. S.*, **22**, 21-5 (2002).
[36] K. Ishikawa, Y. Miyamoto, T. Yuasa, A. Ito, M. Nagayama, and K. Suzuki, "Fabrication of Zn containing apatite cement and its initial evaluation using human osteoblastic cells," *Biomaterials*, **23**, 423-8 (2002).
[37] Y. Sogo, T. Sakurai, K. Onuma, and A. Ito, "The most appropriate (Ca plus Zn)/P molar ratio to minimize the zinc content of ZnTCP/HAP ceramic used in the promotion of bone formation," *J. Biomed. Mater. Res.*, **62**, 457-63 (2002).
[38] H. Kawamura, A. Ito, T. Muramatsu, S. Miyakawa, N. Ochiai, and T. Tateishi, "Long-term implantation of zinc-releasing calcium phosphate ceramics in rabbit femora," *J. Biomed. Mater. Res.*, **65A**, 468-74 (2003).
[39] M. Ikeuchi, A. Ito, Y. Dohi, H. Ohgushi, H. Shimaoka, K. Yonemasu, and T. Tateishi, "Osteogenic differentiation of cultured rat and human bone marrow cells on the surface of zinc-releasing calcium phosphate ceramics," *J. Biomed. Mater. Res.*, **67A**, 1115-22 (2003).
[40] Y. Sogo, A. Ito, M. Kamo, T. Sakurai, K. Onuma, N. Ichinose, M. Otsuka, and R.Z. LeGeros, "Hydrolysis and cytocompatibility of zinc-containing alpha-tricalcium phosphate powder ," *Mat. Sci. Eng. C-Bio. S.*, **24**, 709-15 (2004).
[41] M. Otsuka, Y. Ohshita, S. Marunaka, Y. Matsuda, A. Ito, N. Ichinose, K. Otsuka, and W.I. Higuchi, "Effect of controlled zinc release on bone mineral density from injectable Zn-containing beta-tricalcium phosphate suspension in zinc-deficient diseased rats," *J. Biomed. Mater. Res.*, **69A**, 552-60 (2004).
[42] A. Ito, M. Otsuka, H. Kawamura, M. Ikeuchi, H. Ohgushi, Y. Sogo, and N. Ichinose, "Zinc-containing tricalcium phosphate and related materials for promoting bone formation," *Curr. Appl. Phys.*, **5**, 402-6 (2005).
[43] E. Bouyer, F. Gitzhofer, and M. I. Boulos, "Morphological study of hydroxyapatite nanocrystal suspension," *J. Mater. Sci.-Mater. M.*, **11**, 523-31 (2000).
[44] K. Yoshida, N. Kondo, H. Kita, M. Mitamura, K. Hashimoto, and Y. Toda, "Effect of Substitutional monovalent and divalent metal ions on mechanical properties of β-tricalcium phosphate," *J. Am. Ceram. Soc.*, **88**, 2315-2318 (2005).
[45] X. Wei, and M. Akinc, "Resorption rate tunable bioceramic: Si&Zn-modified tircalcium phosphate," *Advances in Bioceramics and Biocomposites*, **26**, 129-36 (2005).
[46] M. Yashima, A. Sakai, T. Kamiyama, and A. Hoshikawa, "Crystal structure analysis of β-tricalcium phosphate $Ca_3(PO_4)_2$ by neutron powder diffraction," *J. Solid State Chem.*, **175**, 272-7 (2003).
[47] I. Gutowska, Z. Machoy, and B. Machalinski, "The role of bivalent metals in hydroxyapatite structures as revealed by molecular modeling with the HyperChem software," *J. Biomed. Mater. Res.*, **75A**, 788-93 (2005).

[48] L.W. Schroeder, B. Dickens, and W.E. Brown, "Crystallographic studies of the role of Mg as a stabilizing impurity in β-tricalcium phosphate: II Refinement of Mg-containing β-tricalcium phosphate," *J.Solid State Chem.*, **22,** 253-62 (1977).

[49] http://www.atcc.org/common/catalog/numSearch/numResults.cfm?atccNum=CRL-12557.

[50] L.C. Baxter, V. Frauchiger, M. Textor, I. ap Gwynn, and R.G. Richards, "Fibroblast and osteoblast adhesion and morphology on calcium phosphate surfaces," *European Cells and Materials*, **4,** 1-17 (2002).

[51] Z. Schwartz, and B.D. Boyan, "Underlying mechanisms of the bone-biomaterial interface," *J.Cell. Biochem.*, **56,** 340-7 (1994).

[52] K. Mustafa, J. Wroblewski, K. Hultenby, L.B. Silva, and K. Arvidson, "Effects of titanium surfaces blasted with TiO2 particles on the initial attachment and altered cytoskeletal morphology of cells derived from huma madibular bone," *Clin. Oral. Impl. Res.*, **11,** 116-28 (2000).

[53] F.B. Bagambisa, H.F. Kappert, and W. Schilli, "Cellular and molecular biological events at the implant interface," *J. Cranio. Maxill. Surg.*, **22,** 12-7 (1994).

[54] I. Degasne, M.F. Basle, V. Demais, G. Hure, M. Lesourd, B. Grolleau, L. Mercier, and D. Chappard, "Effects of roughness, fibronectin and vitronectin on attachment, spreading and proliferation of human osteoblast-like cells (Saos-2) on titanium surfaces," *Calcif. Tissue Int.*, **64,** 499-507 (1999).

[55] Z. Schwartz, C.H. Lohmann, J. Oefinger, L.F. Bonewald, D.D. Dean, and B.D. Boyan, "Implant surface characteristics modulate differential behaviour of cells in the osteoblast lineage," *Adv. Dent. Res.*, **13,** 38-48 (1999).

[56] H.W. Kim, G. Georgiou, J.C. Knowles, Y.H. Koh, and H.E. Kim, "Calcium phosphates and glass composite coatings on zirconia for enhanced biocompatibility," *Biomaterials*, **25,** 4203-13 (2004).

DETERMINATION OF THE SPATIAL RESOLUTION OF MICRO-FOCUS X-RAY CT SYSTEM WITH A STANDARD SPECIMEN

Mineo Mizuno, Yasutoshi Mizuta, Takeharu Kato and Yasushi Ikeda
Japan Fine Ceramics Center
2-4-1 Mutsuno, Atsuta, Nagoya, 456-8587 Japan

ABSTRACT

A standard specimen for verifying a spatial resolution of three-dimensional images obtained by X-ray computer tomography (XCT) system was developed to analyze in vivo behavior of implant materials. The standard material was made of a single crystal silicon block. There were 4 different kinds of dents on a surface of the block. These dents were dug in by focused ion beam (FIB) etching method. The shape of the dents was square with sides from 1 to 5 μm long. The height of the dents was about 5 μm. Two-dimensional and three-dimensional images, obtained by a micro-focus XCT system, showed the existence of square dents with side 2 μm long and more in the standard specimen. Therefore the spatial resolution of the XCT system was determined to be about 2 μm. The developed standard specimen was proved to be useful for verifying the resolution with XCT systems.

INTRODUCTION

With increasing of the number of the aged, bone disease such as osteoporosis and osteomyelitis are increasing, resulting that bone-substitutes with better qualities are expected to be developed. These materials usually have micro-porous and tailored microscopic structures, because ingrowth ability of new bone is very important for bone regeneration, and the ability has a close relationship with the three-dimensional (3-D) structure[1].

On the other hand, in order to evaluate and analyze bone phenomena and behavior relating to osteoporosis and bone regeneration effects after dosing medicines or implanting biomaterials, in vivo images of 3-D microstructures of bones will be of a great help[2, 3]. A fine and high resolution XCT system is needed[4] to observe a sponge bone with complicated trabecular structures.

Scanning electron microscopy (SEM) is a conventional and popular method to obtain precise 2-D imaging data, but the 3-D images of implant materials can not be obtained[5] by SEM. X-ray computer tomography (XCT) can analyze 3-D image data and is a powerful means to evaluate in vivo behavior of implant materials consecutively and nondestructively[6]. However conventional micro-focus XCT for medical application had the main drawback, because their spatial resolution of about 10 μm and more is too low to obtain high quality images of fine 3-D structures. The images with higher spatial resolution are indispensable to analyze precisely absorption of implant biomaterials and new bone formation in vivo[7, 8]. The current HRXCT, consisting of a micro-focus X-ray tube with a high-precision computing system for 3-D reconstruction and analysis, is expected to have a high spatial resolution for medical application.

On the other hand, it is noted that the high resolution of micro-focus X-ray CT is not easily verified because the standard specimen is not available. Although conventional standard specimens are specified as an ASTM standard[9], a resolution chart or a penetrameter for radiography is a test specimen with many slit lines or holes on a metallic sheet for 2-D images. They are not suitable to 3-D images with high resolution.

Therefore a standard specimen for verifying high spatial resolution of 3-D images is necessary to determine how fine the resolution of a XCT system is.

EXPERIMENTAL PROCEDURE

Preparation of a standard specimen for verifying CT spatial resolution

The material for a standard specimen is a single crystal silicon block. Figure 1 shows a ground plan for the design of the standard specimen, having several square dents with 1, 2, 3 and 5 μm in size. There are five 1 μm-square dents, three 2μm-square dents, three 3μm-square dents, and two 5μm-square dents.

Figure 2 shows the fabrication procedure of the standard specimen. At first, silicon crystal was cut and machined to obtain a rectangular parallelepiped with 1 mm by 1.5 mm by 1.5 mm in size. Then a square-shaped groove with 25 μm in width and 20 μm in depth was grooved on one surface of the block by $Nd/YA_3A_{15}O_{12}$-type laser beam (Laser beam apparatus; Model HCL-2100, HOYA Co., Tokyo, Japan). A square pillar was formed and the area on the top was about 20 μm by 20 μm in size.

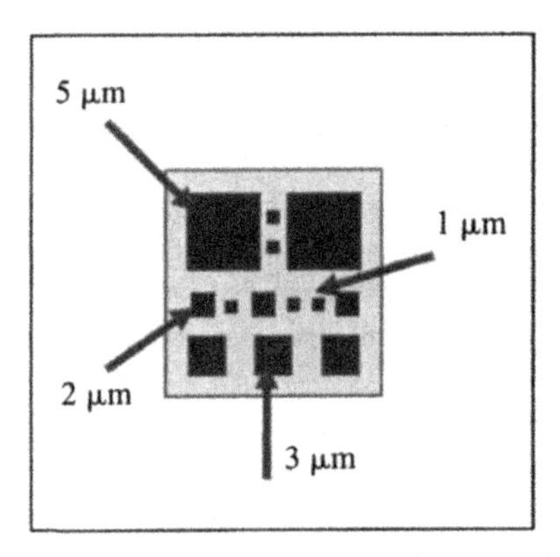

Fig. 1 Design of a standard specimen

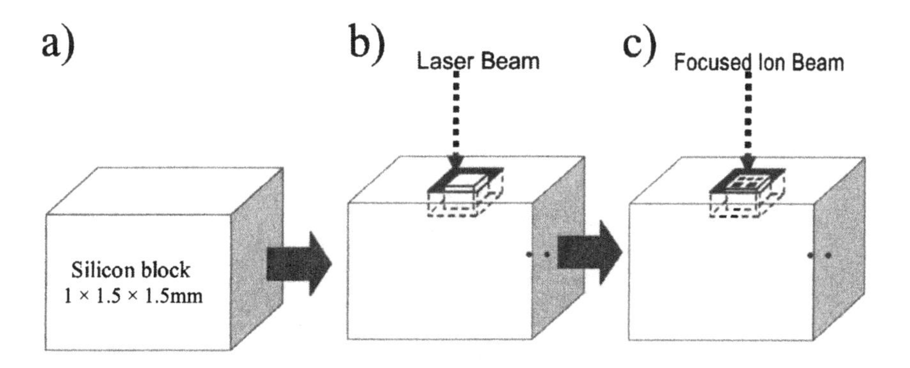

Fig. 2 Fabrication process of a standard specimen
a) Cut and machining
b) Grooving by laser beam
c) Digging in by focused ion beam

Focused ion beam (FIB) etching is a popular method for a specimen preparation of transmission electron microscopy (TEM). As shown in Figure 2, microscopic square dents with sides 1, 2, 3, 5 μm long were dug by focused ion beam (FIB apparatus; Model FB-2100, Hitachi Corp., Japan). The acceleration voltage was 40 KeV. The dent depths were about 5 μm.

Figure 3a shows a scanning electron microscopic image of the standard specimen perpendicular to the LB and FIB etching directions. Figure 3b shows an enlarged image of the dents from slanting position. There clearly observed square dents with 1, 2, 3 and 5 μm in size. The wall surfaces of each dent were smooth and plain. The silicon block was successfully machined according to the design plan as shown in Figure 1. This specimen is found to be suitable to the standard specimen to verify the spatial resolution for HRXCT.

Micro-focus X-ray CT system

Figure 4 shows a scheme of the HRXCT imaging system in this study. The micro-focus generator was operated with the tube voltage of 70 kV, and the electron beam voltage is almost the same as the tube voltage. The electron beam current was 50 μA. The image detector of the CT system is a combined one of a 4-inch type image intensifier and an industrial mono-chronic CCD (130 million pixels type).

The CT projection data was obtained by integrating the 10 real-time projection images, which were the images of 1 sec integration. The number of CT view is 360 or 720, and time for one CT imaging was about 6 or 12 minutes.

The proper dynamic range of the detector system is expected to be more than 1000, but it may be about 256 after 8 bit image processing was carried out.

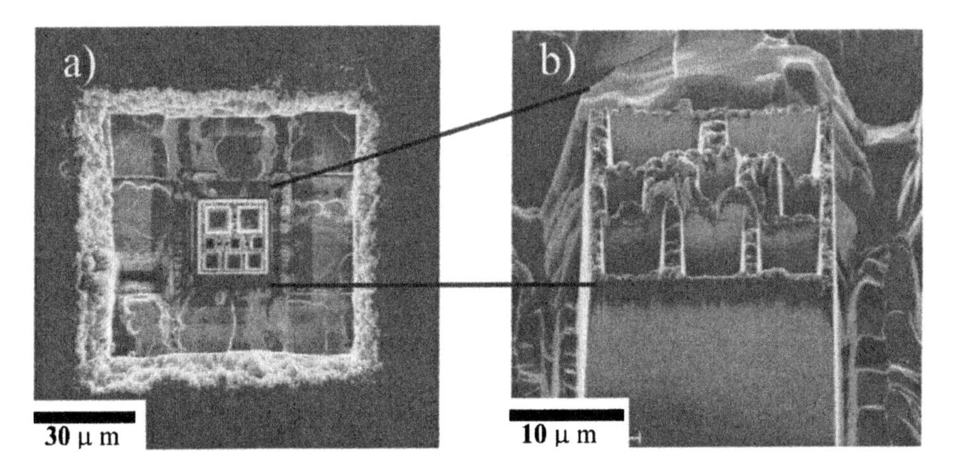

Fig. 3 SEM images of a standard specimen
a) Top image perpendicular to dent
b) Enlarged image of dent from slanting direction

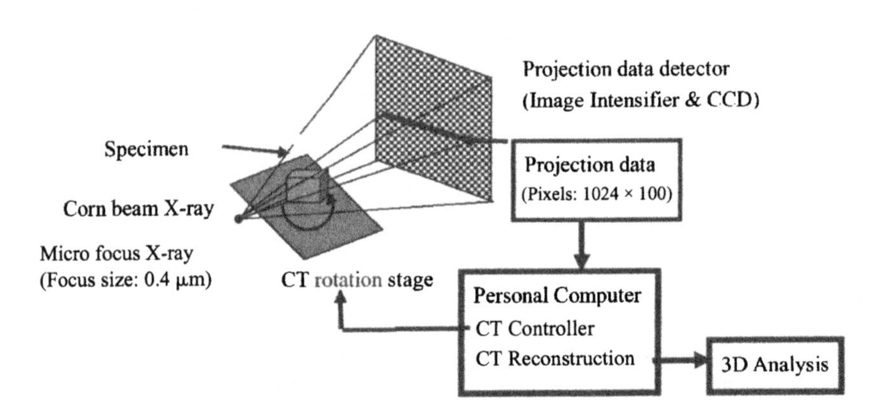

Fig. 4 Set-up of the HRXCT system used in this study

Mechanical oscillation of HRXCT system was depressed as low as possible by depression of oscillation from both vacuum pumps for an X-ray generator and a ground for CT system installation. The X-ray generator and sample stage of CT were combined together, so that the effect of the mechanical oscillation was reduced.

The spatial resolution of XCT mainly depends on a focus size of an X-ray generating point. Figure 5 shows geometrical relationship between a focus size and penumbra (unsharpness of image) size. The special resolution becomes higher with decreasing the focus size. In the case that a focus size is small enough comparing with a defect size in an object, little penumbra is observed as shown in Figure 5. Actually the focus size is not small enough, and the penumbra size (U) on the X-ray projection image of CT is theoretically determined by equation (1).

$$U = (M-1)d \quad (1)$$

where, M is magnification of the imaging ratio of FID (distance between focus and image) to FOD (distance between focus and object), and d is a focus size.

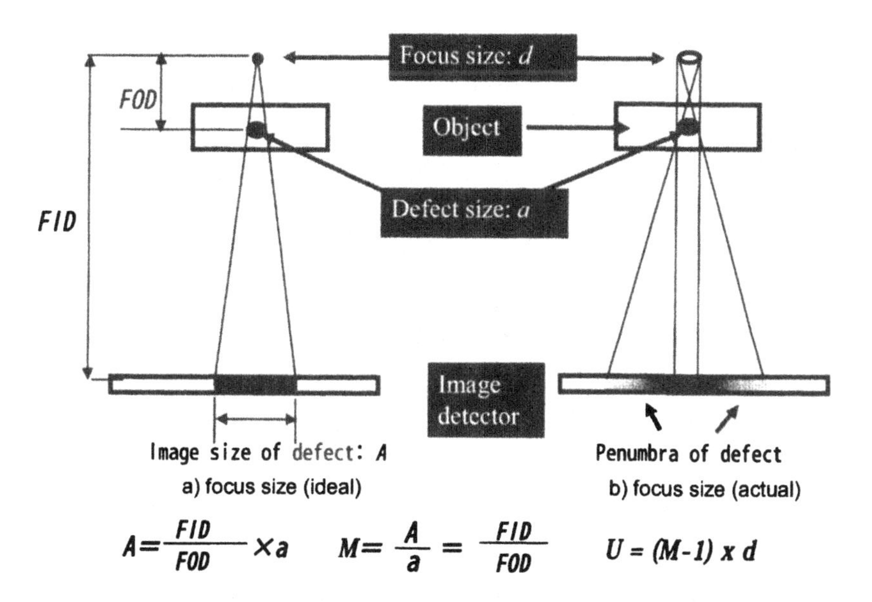

Fig.5 Relationship between focus size and unsharpness of image
U; Penumbra size, A; Image size of defect, d; Focus size
FID; Distance between focus and image,
FOD; Distance between focus and object,

With increasing M, the penumbra size, U, increases linearly. This effect of the magnification on the penumbra is remarkable when the focus size is significantly large. The smaller focus size is important for high resolution observation. In order to obtain HRXCT images with the highest resolution, penumbra size, U, shall be less than a conventional pixel size (=about 100 μm) of the image detector of CT system. In the case that M is 200, the focus size, d, must be smaller than 0.5 μm to make U less than 100 μm. Therefore a focus size of 0.4 μm was used for the HRXCT imaging system in this study (Figure 4).

RESULTS AND DISCUSSION

As shown Figures 4 and 6a, HRXCT measurement was carried out using the standard specimen and the HRXCT system mentioned above. The two-dimensional profile image of the standard specimen is shown in Figures 6b and 6c. The three-dimensional image, reconstructed from the 2-D images, is shown in Figure 7.

Two 5μm-square dents and three 3μm-square dents were distinguished each other in Figure 7. Three 2μm-square dents were also distinguished although the images were not as clear as those from 5μm-square and 3μm-square dents. Five 1 μm-square dents were not identified in both Figures 7 and 6c.

The proper spatial-resolution of the image detector is about 80 μm. One pixel size of the CT was set to be nearly equal to the spatial resolution of the detector. By using the magnification of imaging was 200, the penumbra (geometrical un-sharpness of image) is theoretically 79.6 μm for 0.4 μm of the X-ray focus size. The value of penumbra is not larger than the pixel size of the system. Therefore it can be said that an effect by the geometrical unsharpness is theoretically negligible for our imaging condition. However, a 1 μm-square dent was not distinguished because of the effect by the unsharpness. This suggests that the other factors such as oscillation and contrast resolution increase the unsharpness.

Based on the obtained results and the analysis, the HRXCT system in this study is confirmed to have a resolution of about 2 μm. The standard material for verifying the resolution with X-ray CT systems is expected to be useful for HRXCT observation and evaluation of 3-D structures of implant materials and so on.

CONCLUSION

A standard specimen for verifying the resolution with the HRXCT systems was developed. The material is a silicon block with microscopic dents with sides 1, 2, 3, and 5 mm long. These dents were precisely grooved and dug in with LB and FIB etching methods. The 2-D and 3-D images of the standard specimen taken by the HRXCT system showed that the HRXCT system used in this work has a resolution of about 2 μm with the magnification of 200. The standard specimen is expected to be used for biomaterials research field and so on.

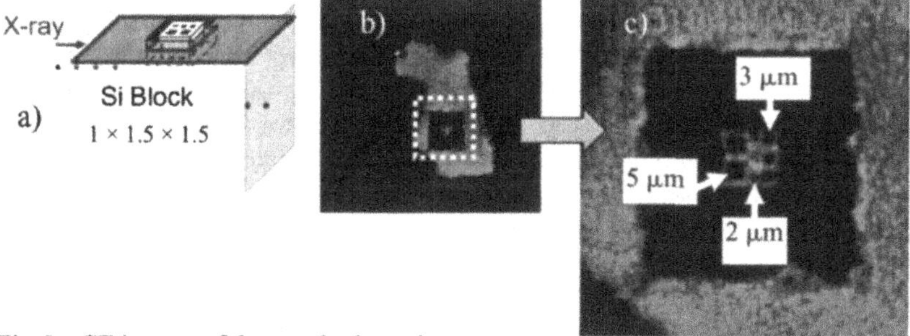

Fig.6 CT images of the standard specimen
a) X-ray direction
b) CT image of engraved area
c) CT image of the main area of standard material

Fig.7 CT images of the standard material

ACKNOWLEDGEMENTS

This research is in part supported by the National Research & Development Programs for Medical and Welfare apparatus entrusted from the New Energy and Industrial Technology Development Organization (NEDO) to the Japan Fine Ceramics Center. The authors thank Messrs. Ryusuke Hirashima and Kazuhito Koizumi of Uni-Hite System Corporation for reconstruction of 3-D images of a standard specimen.

REFERENCES

[1]E. Shors and R. Holmes," Porous Hydroxyapatite," pp.181-99 in *An Introduction to Bioceramics*, edited by L. Hench and J. Wilson, World Scientific Publishing, 1993.

[2]L. Pothuaud, A. Laib, P. Levitz, C. Benhamou and S. Majumdar," Three-Dimensional-Line Skeleton Gragh Analysis of High-Resolution Magnetic Resonance Images: A Validation Study from 34-μm-Resolution Microcomputed Tomography," *J. Bone & Mineral Res.*, **17**, 1883-95 (2002).

[3]A. Laib, H. Hauselmann and P. Ruegsegger," In vivo High Resolution 3D-QCT of Human Forearm," *Technology and Health Care*, **6**, 329-37 (1998).

[4]M. Kothari, T. Keaveny, J. Lin, D. Newitt, H. Genant and S. Majumdar," Impact of Spatial Resolution on the Prediction of Trabecular Architecture Parameters," *Bone*, **22**, 437-43 (1998).

[5]S. Ho, and D. Hutmacher," A Comparison of Micro CT with Other Techniques Used in the Characterization of Scaffolds," *Biomaterials*, **27**, 1362-76 (2006).

[6]M. Ito, T. Nakamura, T. Matsumoto, K. Tsurusaki and K. Hayashi," Analysis of Trabecular Microarchitecture of Human Iliac Bone Using Microcomputed Tomography in Patients with Hip Arthrosis with or without Vertebral Fracture," *Bone*, **23**, 163-69, (1998)..

[7]Y. Ikeda, M. Yasutomi, M. Mizuno, M. Mukaida, M. Neo and T. Nakamura," 3 Dimensional CT Analyses of Bone Formation in Porous Ceramic Biomaterials," pp.185-90 in Ceram. Eng. & Sci. Proceedings Vol.24, Issue 3, Am. Ceram. Soc., 2003.

[8]Y. Ikeda, Y. Mizuta, M. Mizuno, K. Ohsawa, M. Neo and T. Nakamura," 3D CT Analyses of Porous Structures of Apatite Ceramics and in vivo Bone Formation," pp.833-38 in Ceram. Eng. & Sci. Proceedings Vol.23, Issue 4, Am. Ceram. Soc., 2002.

[9]ASTM E 747, "Standard Test method for Controlling Quality of Radiographic Examination Using Wire Penetrameters," ASTM, 1990.

Processing of Biomaterials

HYDROXYAPATITE HYBRIDIZED WITH METAL OXIDES FOR BIOMEDICAL APPLICATIONS

Akiyoshi Osaka[1,2]*, Eiji Fujii[3], Koji Kawabata[3], Hideyuki Yoshimatsu[3], Satoshi Hayakawa[1], Kanji Tsuru[1,3], Christian Bonhomme[4], Florence Babonneau[4]

[1] Faculty of Engineering, Okayama University
Tsushima, Okayama, 700-8530, Japan
[2] Research Center for Biomedical Engineering, Okayama University
Tsushima, Okayama, 700-8530, Japan
[3] Industrial Technology Center of Okayama Prefecture
Haga, Okayama, 701-1296, Japan
[4] Laboratoire de Chimie de la Matire Condense de Paris,
Universite Pierre et Marie Curie, CNRS
Jussieu, 75252 Paris Cedex 05, France

ABSTRACT

Zn-substituted apatites with homogeneous Zn distribution in the particles (Zn•HAp) and those with graded distribution (Zn/HAp) were prepared via a conventional wet chemical synthesis and soaking the Zn-free apatite particles in zinc nitrate solutions: X-ray photoelectron spectra of the apatites confirmed such Zn distribution. Physical and chemical properties such as particle size, specific surface area, or zeta-potential and were correlated to adsorption of β_2-microglobulin (β_2-MG) and bovine serum albumin (BSA). Adsorption of β_2-MG on both apatites increased with the Zn content, whereas more β_2-MG was adsorbed on Zn/HAp than on Zn•HAp. The amount of BSA adsorbed on Zn•HAp reached a maximum at about 1.5%, while it remained constant for Zn/HAp. From the results, it was suggested that Zn-substituted sites had better affinity against β_2-MG and favored β_2-MG adsorption. Surface charge on the particles was one of the factors to explain the suppressed adsorption of BSA by both apatites.

1. INTRODUCTION

Recently, adsorbents of pathogenic substances without adsorbing essential proteins have been demanded. Among those substances, β_2-microglobulin (β_2-MG)[1] is a precursor protein of amyloid that causes dialysis-related amyloidosis and is not removed by the current dialysis therapy. All proteins in our blood are more or less charged, and, ceramic particles also are charged when dispersed in solutions. Hence, one may take advantage of the electrical charge distribution to distinguish and separate such pathogenic substances from essential proteins. In addition, most ceramic particles are secondary ones, consisting of agglomeration of relevant primary particles. Then, the size of interstitial voids or spaces among the agglomerations may vary due to the size. When optimum conditions are attained, some molecules are stacked in them.

Hydroxyapatite ($Ca_5(PO_4)_3OH$ in stoichiometric composition; HAp) and its derivatives are widely used in biomedical fields due to their biocompatibility and physiological activity, such as cell attachment, stimulating cell proliferation and differentiation, or protein and enzyme adsorption[2]. It is common for apatites that substitution of the component ions varies the primary particle size, crystallinity, and surface charge. Thus, one may obtain such apatites that exhibit optimized selectivity in adsorbing a specific pathogenic substance due to cation or anion substitution[3)-6)]. Si(IV) is by far harmless but a preliminary study by Fujii et al.[7)] indicated that the amount of substitution in the apatite lattice might be minimum as Al(III) or Ti(IV) is impractical. Fujii et al.[7)] confirmed that Zn^{2+} enters the apatite lattice, though it seems strange when the difference in the ionic size of Zn^{2+} and Ca^{2+} is considered. Moreover, Miyaji et al.[8)] found decrease in the apatite lattice due to the increase in the Zn^{2+} content. Thus, the incorporation of Zn^{2+} substituting Ca^{2+} in the apatite lattice seems acceptable.

The presence of Ca^{2+} ion channels and ion-exchange via the channels are two of the characteristic aspects of apatite. The substituting ions can enter the lattice through the channels to occupy the Ca^{2+} sites. Therefore, one may employ at least two routs to replace Ca^{2+} ions with Zn^{2+} ions. One is an ordinary rout in which a soluble zinc compound is mixed together with a Ca source in the same starting solution: the other one is to conduct ion-exchange where apatite particles derived in any way are to be soaked in a Zn^{2+} solution. The Zn^{2+} ions are mostly homogeneously distributed in the apatite particles derived from the former rout. In contrast, in the latter particles, they may only reside on the surface layer or are distributed with a degraded concentration profile. As adsorption is a physico-chemical process to take place at the surface, the latter particles seem more practical with respect to save total Zn content in the particles as far as other physical and chemical characteristics remain similar to the particles with homogeneous distribution. One of the aims of the present paper is to of the compare those particles in terms of particle size (primary and secondary), surface charges or zeta-potential, and selectivity in adsorbing ß$_2$-MG over bovine serum albumin (BSA). Ergun et al.[9)] indicated that Zn-containing apatite adsorbed less amount of BSA than Zn-free apatite. It is suggested, then, that the Zn-substituted apatites are good candidates for separation of ß$_2$-MG when they adsorb well ß$_2$-MG.

2. EXPERIMENTAL

Hydroxyapatite and Zn-substituted apatites were synthesized from reagent-grade calcium nitrate, zinc nitrate, and diammonium hydrogen phosphate due to a wet chemical method. The phosphate aqueous solution (0.3 mol/L, 400 mL) whose pH was adjusted to 10 by addition of a 28 mass% NH_4OH aqueous solution was added under rigorous stirring to a Ca-Zn mixed solution (400 mL) at a feeding rate of 3 mL/min under an N_2 atmosphere at 60°C. Here, the mixed solutions were prepared by adding 0.5 mol/L zinc nitrate solution to the 0.5 mol/L calcium nitrate solution so that Zn content in the precursor solutions for apatites was in the range 0 to 0.05 mol/L. After completion of the addition, the precipitates were aged for 24h, washed with distilled water, and dried at 105°C for 48h. The derived cakes were milled and sieved to obtain particles of <150 µm in size. The Zn^{2+} ions were presumably distributed homogeneously in those apatites. They were denoted as Zn•HAp.

The same procedure was taken for preparing Zn-free hydroxyapatite. Thus-obtained HAp particles (10 g) were dispersed in 400 mL of 0-11.5mM $Zn(NO_3)_2$ aqueous solution. The system was aged at 80°C for 24h, before the particles were filtrated, washed with distilled water and dried at 105°C for 48h. Those samples were denoted as Zn/HAp.

Zn(II), Ca(II), and P(V) contents in both apatites were analyzed by the inductively coupled plasma emission spectroscopy (ICP, ICPS-7500, SHIMADZU). The Zn content in the both apatites (Zn•HAp's and Zn/HAp's) were expressed in mass %, which enables to compare the behavior of both series of apatites that were different in the Zn distribution (homogeneous and graded). The crystalline phases were identified with an X-ray diffractometer (XRD, RINT2000, RIGAKU; 40 kV-200 mA), and the lattice parameters were derived using the software involved in the system-controlling unit. The specific surface area (SSA) were measured using the BET N_2 adsorption method (GEMINI2370 Micromeritics⌋SHIMADZU). Morphology of those apatite particles were observed by transmission electron microscopy (TEM, JEM-2010⊐JEOL). Zeta potential was measured with a ZETASIZER machine (ZETASIZER 3000HS_A, MALVERN), using physiological saline (0.142 mol/L NaCl aqueous solution, 7.4 in pH) as the dispersion medium. X-ray photoelectron spectra of the Zn $2p_{1/2}$ core level for those apatites were measured on a S-Probe ESCA SSX 100S machine (Fisons Instruments, UK) with monochromated Al Kα irradiation. Detail of similar measuring procedure or conditions was presented elsewhere.[10)]

Both BSA and ß$_2$-MG were dissolved in the saline to prepare mixed-protein solution so as to attain the concentration of BSA to be 70 mg/mL and that of ß$_2$-MG to be 30 μg/mL. Note here that this mixed protein solution simulates the blood plasma of people suffering from amiloidosis. The apatite powders, 0.1 g of Zn•HAp and Zn/HAp (<150μm), were dispersed first in 1.0 mL of the saline, to which 1.0 mL of the mixed protein solution was added to be contacted for 6 h. Thus, the actual concentration of the BSA in the solutions in contact with the sample powders was 35 mg/mL and that of ß$_2$-MG was 15 μm/mL. The amounts of BSA and ß$_2$-MG adsorbed on the apatite particles were quantitatively analyzed by colorimetry, taking the standard procedure whose detail was described elsewhere.[6)].

3. RESULTS

Figures 1 (a) and (b) show the XRD patterns of Zn•HAp's and Zn/HAp's, respectively. All peaks were assigned to hydroxyapatite (JCPDS: 09-0432), while no peaks were detected for possible byproducts such as $CaZn_2(PO_4)_2 \cdot 2H_2O$ (JCPDS: 35-0495), $Zn_2P_2O_7 \cdot 5H_2O$ (JCPDS:07-0087), ZnO (JCPDS: 21-1486, 36-1451). The profile for Zn•HAp's lost sharpness with increased Zn content, indicating the decrease in crystallinity or particle size. In contrast, the XRD profile for Zn/HAp's remained sharp. Thus, with the width of (002) diffraction at ~26° and Scherer equation, the crystallite size was estimated: as expected, the size for Zn•HAp's decreased almost linearly with the Zn content from ~28 nm for 0.3% to 15nm for 5.9% while it was ~43 nm for Zn-free HAp. The crystallite size for the Zn/HAp was almost independent of the Zn content: it was about 50 nm. The estimation agrees with the transmission electron micrographs, given in Figs. 2 and 3 for Zn•HAp's and Zn/HAp's. Here, the particle size of Zn•HAp's decreased from ~40 nm (0.9 mass%) to ~20 nm (5 mass%), while that of Zn/HAp's was about 50 nm and scarcely dependent upon the Zn content.

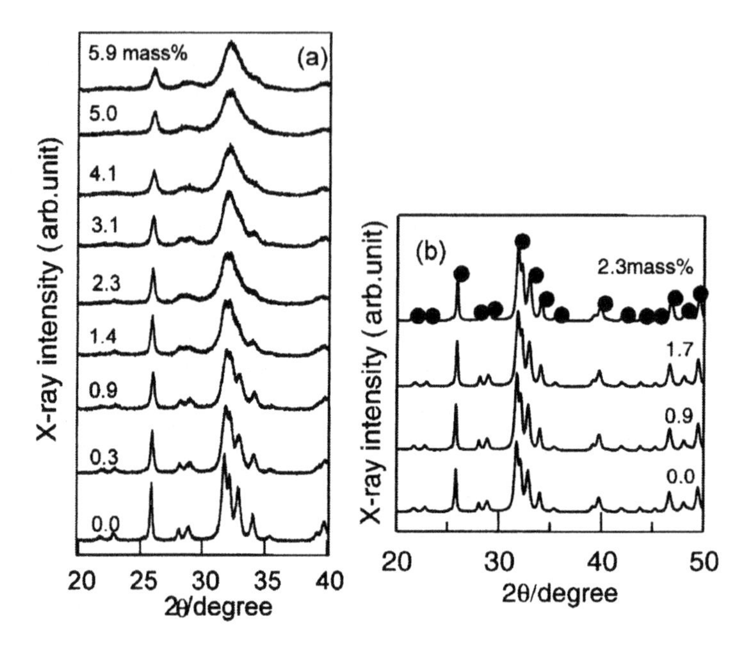

Fig. 1 XRD patterns for (a) Zn•HAp and (b) Zn/HAp powders. The figures represent the Zn content in mass%. All peaks are assignable to apatite (JCPDS 09-0432), as indicated in the top profile in (b) with ●.

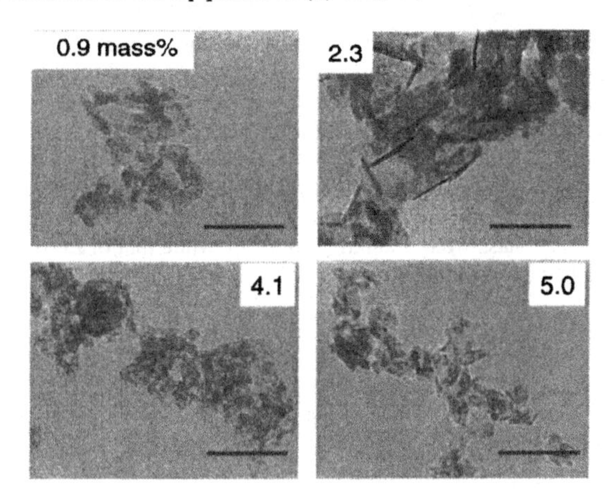

Fig. 2 TEM photographs for selected Zn•HAp particles. The figures are the Zn contents in mass%. Bar: 100nm

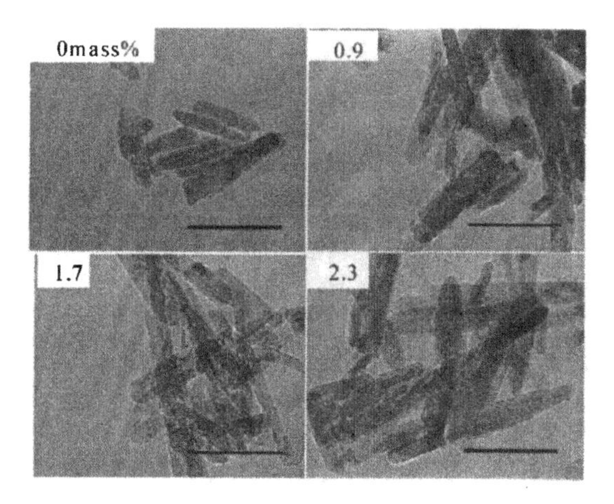

Fig. 3 TEM photographs of Zn/HAp particles. The figures represent the Zn content in mass%. Bar: 100nm

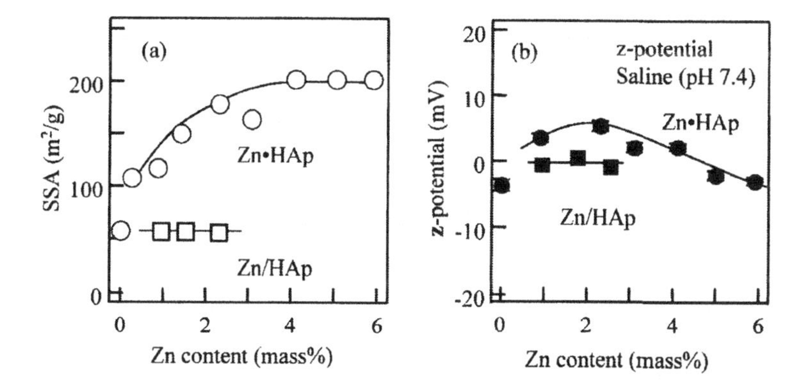

Fig. 4 (a) SSA and (b) z-potential for Zn•HAp and Zn/HAp as a function of the Zn content (mass%).

Fig. 4(a) indicates that, in accordance with the particle size changes, SSA for Zn•HAp increased with the Zn content, while that for Zn/HAp remained almost constant. Fig. 4 (b) compares zeta-potential between Zn•HAp's and Zn/HAp's. With involving very small amount of Zn, the zeta-potential for Zn•HAp's jumped to a positive value from a negative value for Zn-free HAp, then gradually decreased below zero beyond 4 mass%. In contrast, that for Zn/HAp was practically constant and was scattered around zero. Those changes will be discussed together with the protein adsorption characteristics presented below.

The intrinsic protein adsorption ability of the apatites is to be compared on the unit area basis, though the weight basis adsorption ability is essential for columnar packages in practical therapeutic applications. In this respect, the amount of BSA and ß$_2$-MG adsorbed per unit area is plotted in Fig. 5 as a function of the Zn content (<3mass%). Fig. 5 (a) indicates that the ß$_2$-MG adsorption on both apatites increased up to 3 mass%: 1.6μg/m^2 for Zn-free HAp, 2.5μg/m^2 for 2.3%Zn/HAp, and 2.3μg/m^2 for 3.1%Zn•HAp. Due to the limitation of ion exchange, no data were available at this moment for Zn/HAp with more Zn contents than 3%. Zn•HAp's involving more Zn were available, but no more ß$_2$-MG than ~2.3μg/m^2 was adsorbed if the Zn content was further increased beyond 3%. It is emphasized that Zn/HAp's adsorbed more ß$_2$-MG than Zn•HAp. In Fig. 5(b) the BSA adsorption on Zn•HAp's increased to reach a maximum (1.5mg/m^2) at about 1.5 mass% and then decreased. With further increase in the Zn content over 4%, Zn•HAp adsorbed little amount of BSA, though not indicated here. Selective adsorption means adsorption of a target protein relative to that of reference proteins. In this regard, the Zn incorporation has improved the selectivity in terms of ß$_2$-MG for both series of apatites.

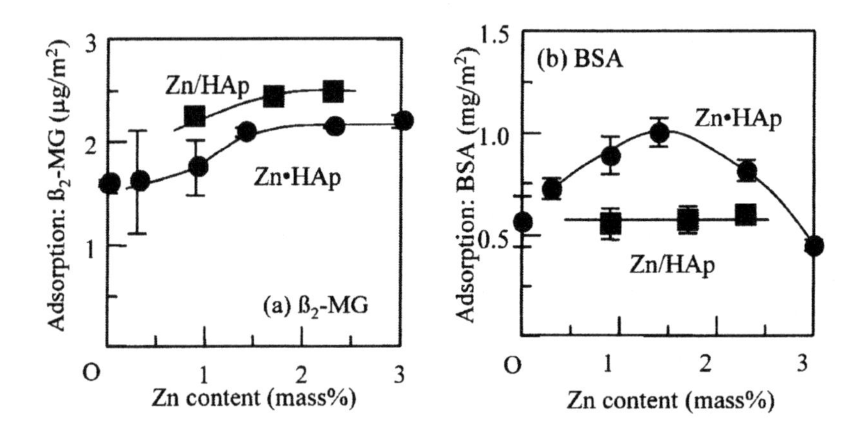

Fig. 5 Protein adsorption on Zn•HAp's and Zn/HAp's: (a) ß$_2$-MG, (b) BSA.

4. DISCUSSION

4.1 Zn substitution and distribution in the apatite particles

Unfortunately, no direct evidence has been present to confirm incorporation of Zn^{2+} ions in the apatite lattice.[8,9,11] The lattice parameters for Zn/HAp's from their XRD data decreased with increase in the Zn content: (a_0, c_0): (0.944 nm, 0.688nm) for Zn-free HAp to (0.938 nm, 0.682nm) for 3.1%Zn/HAp. The fraction of the decrease, about 0.6~0.8%, has been considered significant when the accuracy of the derivation is estimated to be <0.2%. Note here that whereas Ergun et al.[9] gave (0.94270, 0.68878) for HAp, and (0.93828, 0.68760) for 2mol%-Zn, their apatites were heated at 1100°C, and thus the data should not be compared with the present ones.

If Zn substitutes Ca in the apatite lattice, it is left unanswered how the small Zn^{2+} ion should stabilize the larger Ca site. Since no other adequate interpretation of the decrease in the lattice parameters, it has been assumed that Zn substitutes Ca as mentioned above in Introduction section, referring to Miyaji et al.[8] Fig. 6 indicates X-ray photoelectron spectra (XPS) of the apatites. XPS represent the information collected from the material surface; hence, the peak intensity shows relative abundance of Zn in the surface shell of the apatite particles. It is evident in the figure that the Zn $2p_{1/2}$ peak intensity for both apatites grew with the Zn content, while the peaks for Zn•HAp were weaker than those for Zn/HAp with similar Zn contents. Thus, it has been concluded that more Zn ions were involved in the Zn•HAp particles when the solutions of larger Zn contents were employed, and, what is more important, more Zn were present on the surface shell layers of the Zn/HAp particles than on Zn•HAp.

4.2 Protein adsorption

Since both proteins are negatively charged in the saline (7.4 in pH), the particles with positive electrical double layer, i.e., positive in zeta-potential, would be favorable in terms of adsorbing either of them. Indeed, Fig. 4(b) and Fig. 5(b) indicate that zeta-potential and BSA

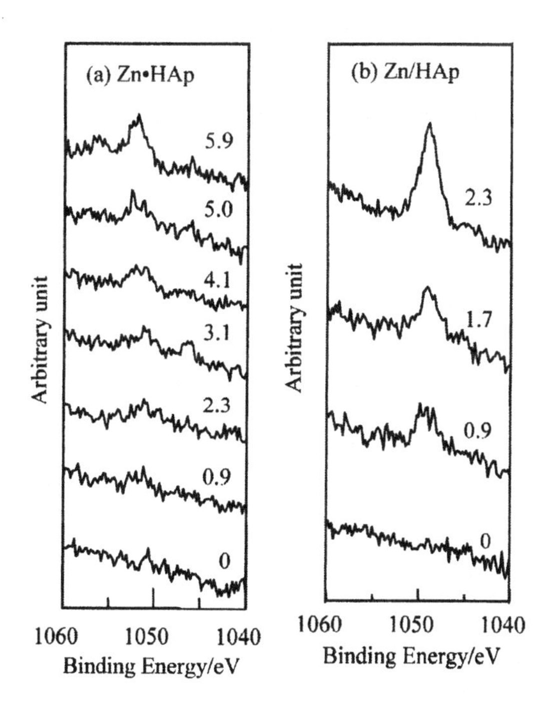

Fig. 6 X-ray photoelectron spectra of Zn $2p_{1/2}$ for the apatites: (a) Zn-Hap (b) Zn/Hap. The figures represent the Zn content in mass%.

adsorption on either of Zn•HAp and Zn/HAp are similarly dependent on the Zn content. Thus, BSA adsorption is explained due to surface charge of the particles. No such similarity, however, has been detected for the dependence of zeta-potential and β_2-MG adsorption on the Zn content. If Zn-substituted apatites could be assumed to have better affinity with β_2-MG, β_2-MG adsorption dependent on the Zn content shown in Fig. 5(a) could be reasonably interpreted, on the basis of the XPS data in Fig. 5: more Zn^{2+} ions were present on the surface shell of Zn/HAp than on that of Zn•HAp, and the Zn content on the surface increased for both apatites. Therefore, the present results strongly suggest such better affinity, or the Zn sites are active preferably to adsorb β_2-MG. Unfortunately, however, no rational explanation could be given on the reason why the Zn-incorporated apatites can distinguish BSA from β_2-MG, and why zeta-potential of Zn/HAp's showed no change despite that the surface Zn content increased. The latter issue could speculatively interpreted as: the Zn concentration of the outermost surface of Zn/HAp was constant, or there is a threshold Zn^{2+} ion content in the apatite, and no more Zn^{2+} ions could be incorporated; apparent increase in the Zn content confirmed by XPS in Fig 6 is due to the increase in the Zn content in the inner part of the surface shell zone near the particle core. Since no data available on the Zn distribution profile in the apatite particles, this interpretation provides only a hypothesis.

Another possible interpretation is due to the interstitial spaces (meso pores) among the particles in the agglomerates. Fujii et al.[11] pointed out that meso pores formed in the agglomerates of Zn•HAp particles were smaller than BSA but were adequate to accommodate β_2-MG. Ando et al.[12] emphasized combination of the pore size and Zn sites lead to better selectivity of β_2-MG for Zn/HAp's.

CONCLUSION

Two series of Zn-substituted apatites, Zn•HAp and Zn/HAp, were prepared. With calcium nitrate, diammonium monohydrogen phosphate, and zinc nitrate as the starting chemicals, a conventional wet chemical procedure was taken for Zn•HAp particles in which Zn ions were homogeneously distributed. Zn/HAp particles that had graded distribution of Zn were prepared by soaking the Zn-free apatite particles in zinc nitrate solution. Both apatites were characterized in terms of particle size, specific surface area, zeta-potential and lattice parameters. The Zn distribution was confirmed by their X-ray photoelectron spectra. In addition, β_2-MG and BSA mixed solutions were contacted with those particles, and adsorption of those proteins on the apatites was measured.

The particle size of Zn•HAp's decreased (40 nm to 20 nm) with the Zn content (0.9 mass % Zn ~ 5 %) while that of Zn/HAp's remained constant (~50 nm). Specific surface area changed accordingly: it increased for Zn•HAp (50 ~ 200 m^2/g), and remained constant for Zn/HAp (~50m^2/g). Zeta-potential for Zn•HAp increased with the Zn content, and reached a maximum (~+5mV) at ~2 mass%. It then decreased to take negative values beyond 4mass%. As expectedly, no dependence of zeta-potential on the Zn content was observed for Zn/HAp. Adsorption of β_2-MG on both apatites increased with the Zn content, whereas more β_2-MG was adsorbed on Zn/HAp than on Zn•HAp. In contrast, differently depended was BSA adsorption: The amount of BSA adsorbed on Zn•HAp reached a maximum at about 1.5%, while it remained constant for Zn/HAp. From the results, it was suggested that Zn-substituted sites had better affinity against β_2-MG and favored β_2-MG adsorption. The decrease in adsorption of BSA by both apatites was attributable to surface charge effects while the pore size could be a factor as pointed out by Fujii et al.[11]

ACKNOWLEDGEMENTS

This study was supported by a Grant-in-Aid for Scientific Research from the Japan Society for the Promotion of Science (No.16360330, Z1416005), and The Hosokawa Powder Technology Foundation.

REFERENCES

1 F. Gejyo, T. Yamada, S. Odani, Y. Nakagawa, M. Ayakawa, T. Kunitomo, H. Kataoka, M. Suzuki, Y. Hirasawa, T. Shirahama, A. Cohen and K. Schmid, "A new form of amyloid protein associated with chronic hemodialysis was identified as $ß_2$-microglobulin," *Biochem. Biophys. Res. Commum.*, **129**, 701-706 (1985).

2 C.A. Zittle, "Adsorption studies of enzymes and other proteins," *Advances in Enzymol*, **14**, 319-374 (1951).

3 S. Takashima, S. Hayakawa, C. Ohtsuki and A. Osaka, "Adsorption of proteins by calcium phosphate with varied Ca to P ratios," *Bioceramics*, **9**, 217-220 (1996).

4 S. Takashima, Y. Kusudo, S. Takemoto, K. Tsuru, S. Hayakawa and A. Osaka, "Synthesis of carbonate-hydroxy apatite and selective adsorption activity against specific pathogenic substances," *Bioceramics*, **14**, 175-178 (2002).

5 S. Hayakawa, Y. Kusudo, S. Takemoto, K. Tsuru and A. Osaka, "Hydroxy-carbonate apatite, blood compatibility and adsorption of specific pathogenic proteins," *Bioceramics:Materials and Applications IV*, **147**, 111-119 (2003).

6 S. Takemoto, Y. Kusudo, K. Tsuru, S. Hayakawa, A. Osaka and S. Takashima, "Selective protein adsorption and blood compatibility of hydroxy-carbonate apatites," *J. Biomed. Mater. Res.*, **69A**, 544-551 (2004).

7 E. Fujii, K. Kawabata, H. Yoshimatsu, S. Hayakawa, K. Tsuru, and A. Osaka, "Properties of metal oxide doped hydroxyapatite powder prepared by sol-gel method" J. Ceram. Soc. Japan Suppl. 112-1 (Special Issue: The 5th International Symposiums of Pacific Rim Ceramic Societies, Nagoya, September 29-October 2, 2004), **112**, S835-S838 (2004).

8 F. Miyaji, Y. Kono and Y. Suyama, "Formation and structure of zinc-substituted calcium hydroxyapatite," *Mater. Res. Bull.*, **40**, 209-220 (2005)

9 C. Ergun, T. J. Webster, R. Bizios, R. H. Doremus, "Hydroxylapatite with substitute magnesium, Zinc, cadmium, and yttrium. I. Structure and microstructure", *J. Biomed. Mater. Res.*, **59**, 305-311 (2002).

10 E. Fujii, K. Kawabata, H. Yoshimatsu, S. Hayakawa, K. Tsuru, A. Osaka, "Structure and Biomineralization of Calcium Silicate Glasses Containing Fluoride Ions," J. Ceram. Soc. Japan, **111**, 762-766 (2003).

11 E. Fujii, M. Ohkubo, S. Hayakawa, K. Tsuru, A. Osaka, K. Kawabata, C. Bonhomme, F. Babonneau, "Selective protein adsorption property and characterization of nano-crystalline zinc-containing hydroxyapatite," *Acta Biomaterialia*, **2**, 69-74 (2006).

12 K. Ando, M. Ohkubo, S. Hayakawa, K. Tsuru, A. Osaka, E. Fujii, K. Kawabata, C. Bonhomme, F. Babonneau, " Synthesis and structural characterization of nanoapatite ceramics powders for biomedical applications," presented at Pacific Rim 6 (The 6th International Symposiums of Pacific Rim Ceramic Societies), Hawaii, September 11-16, 2005.

PREPARATION OF SELF-SETTING CEMENT-BASED MICRO- AND MACROPOROUS GRANULES OF CARBONATED APATITIC CALCIUM PHOSPHATE

A. Cuneyt Tas
Clemson University
School of Materials Science and Engineering
Clemson, SC 29634

ABSTRACT

A method of preparing spherical, micro- and macroporous (50 to 550 μm pores), carbonated apatitic calcium phosphate granules (2 to 4 mm in size) has been developed by using the NaCl porogen technique. A calcium phosphate self-setting cement powder (comprising a specially designed cement powder mixture of α-$Ca_3(PO_4)_2$, $CaHPO_4$, $CaCO_3$ and precipitated $Ca_{10}(PO_4)_6(OH)_2$) was mixed with 65 wt% NaCl crystals ranging in size from 1 mm down to 400 μm, followed by kneading with a dilute Na_2HPO_4 initiator solution and then sieving the setting paste to the desired sizes. Embedded NaCl crystals were leached out from the formed granules by soaking in deionized water at room temperature. The calcium phosphate granules comprised macro- and micropores substantially communicating with one another throughout the body with a porosity of 45-50% or more. Produced granules were only composed of carbonated, calcium-deficient, poorly-crystallized, apatitic calcium phosphate as the mineral phase, which is quite similar to that of human bones. Granules are used (marketed in Europe under the trade name of "Calcibon® Granules") as a substitute or a repair material for bone, carrier material for drug delivery and controlled release system. These granules have been the first calcium phosphate granules directly produced from a self-setting calcium phosphate orthopedic cement powder at room temperature, and they are also suitable for augmentation with autologous bone graft, bone marrow aspirate, blood or platelet-rich plasma.

INTRODUCTION

Calcium phosphate (CaP) bioceramics are used for bone reconstruction because of their close resemblance to the bone mineral phase, i.e., biological apatite. Stoichiometric calcium hydroxyapatite (HA: $Ca_{10}(PO_4)_6(OH)_2$) is far from mimicking the bone mineral. "Biological" or "bone-like" apatites, which constitute the mineralized portion of bones, are carbonated (4 to 6 wt%, and this fact alone makes them somewhat closer to the mineral dahllite rather than hydroxyapatite), poorly-crystallized, alkali (i.e., Na and K) and alkaline earth (Mg) element-doped, non-stoichiometric, calcium-deficient (about 10%) apatitic phosphates with a Ca/P molar ratio variable over the range of 1.55 to 1.70 [1-6]. In addition to possessing a chemical make-up closer to that of natural bone, orthopedic implants should have particularly designed structural characteristics, in order to better serve as desired. Orthopedic implants developed for bone repair should be porous, so as to invite ingrowth of newly formed bone into the implant, leading to a more securely fixed and integrated repair. Porous structures are particularly favorable when utilized in conjunction with natural cancellous bone, as they can closely mirror the structure of the host bone.

Bioactive and biocompatible CaP ceramics exhibit excellent osteoconductive properties. CaP-based bone substitutes are typically used for bone replacement or augmentation in a wide spectrum of clinical applications [7-9]. CaP ceramics used as bone substitutes are commercially available as single-phase powders, self-setting cements [10-19], granules [20-23] or macroporous blocks [24-27]. Synthetic CaP bone substitutes with interconnected macropores will facilitate the penetration of cells and biologic growth factors into the implant allowing the osteogenic process to occur within the inner surfaces of pores. Macropores (from 100 to 600 μm), as well as micropores (i.e., pores ranging from 3 to 30 μm), are shown to be necessary for bone ingrowth to take place in synthetic bone substitute materials [28, 29].

Bone substitutes in the shape of granules of well-defined geometry, preferably in the shape of spheres in different sizes, provide the surgeon with an unmatched ease in filling the bone defects of irregular shape, in comparison to pre-shaped prismatic blocks. Granules, impregnated with, for instance, platelet-rich plasma, will be easily packed together by the orthopedic surgeon to fill and reconstruct the bone voids that may have been caused by trauma or other genesis, such as a benign tumor, from surgery or congenital defects.

Over the last two decades, hydroxyapatite-based granules have received quite a significant attention from both the orthopedic surgeons and the materials scientists, and consequently, numerous reports were encountered in the open literature [30-72]. Highlights of the previous research on CaP granule production will be briefly summarized in the following. Fabbri *et al.* [39] formed millimeter-sized HA granules by dripping a ceramic suspension into liquid nitrogen, followed by sintering. Granules of coralline-origin materials were prepared and tested *in vivo* by Holtgrave [41] and Baran *et al* [71]. Spray granulation process has also been tested for the micron-sized CaP granule production [42, 47]. However, this technique also necessitated a follow-up sintering step. Liu *et al.* [44] modified the dripping procedure originally developed by Fabbri *et al* [39], and added polyvinylbutyrale (PVB) into the ceramic suspensions. Granules were then sintered at 1200°C. The use of PVB, as the porogen, was also tested by Zyman *et al.* [72] in forming porous granules. Complete burn off of PVB always necessitated high temperatures ($<$1150°C). Maruyama *et al.* [46] formed a paste consisting of HA powders, CaO, ZnO, chitosan and malic acid, wet granules were first formed out of this paste, followed by sintering at 1150°C. Gauthier *et al.* [50] used the technique of wet granulation on chemically synthesized CaP powders to obtain 200 to 500 microns granules. Oonishi *et al.* [22, 56] tested the *in vivo* response of granules made from poorly-crystallized CaP powder mixtures. Merkx *et al.* [51] examined the *in vivo* behavior of porous, bovine-origin granules in a goat model. Paul *et al.* [54] first prepared the HA powders by starting with $Ca(OH)_2$ and H_3PO_4, then formed viscous suspensions by using chitosan and paraffin, and obtained granules upon stirring in those thick suspensions. Formed granules were fired at 1100°C to achieve the strength required for handling. Patel *et al.* [66] used a similar procedure to produce CaP granules with that used by Paul *et al* [54]. Preparation of apatite-wollastonite (A-W) glass-ceramic granules were reported by Ikeda *et al.* [53]. Schwartz *et al.* [55] studied the preparation of porous biphasic (HA-$Ca_3(PO_4)_2$) granules after sintering the wet, organic substance-containing granulates at 1200°C. A similar procedure for granule manufacture was recently reported by Tanaka *et al* [69]. Barinov *et al.* [63-65] first prepared a CaP suspension, and then dissolved gelatin in it. Upon stirring this suspension in a bath of oil, spherical granules of micron size were formed. Wet granules were collected and then sintered. Rodriguez-Lorenzo *et al.* [67] prepared HA suspensions containing soluble starch, and utilized the swelling of starch to form pores, followed by sintering at 1100°C and granulation. Takagi *et al.* [61] blended CaP powders with sucrose and $NaHCO_3$, followed by granulation, to obtain porous granules with sizes ranging from 125 to 250 microns.

However, none of the CaP granules used or manufactured in these previous studies [30-72] was able to simultaneously meet the following crucial criteria for a successful porous bone substitute:

(1) to be produced without being heated at temperature higher than the physiologic temperature, and without sintering,
(2) to be comprised of carbonated, calcium-deficient, apatitic CaP just as the human bones, not just of stoichiometric HA or TCP ($Ca_3(PO_4)_2$) ceramic,
(3) to possess an interconnected network of micro- and macropores,
(4) to have the ability of being impregnated (i.e., wicking ability) with the patient's own blood or bone marrow aspirate prior to the implantation,
(5) to contain nanocrystals of carbonated, calcium-deficient apatitic CaP on its surface.

This study [73] reported a simple, industrial method of preparing micro- and macroporous, Ca-deficient, carbonated, apatitic calcium phosphate granules over the size range of 2 to 4 mm, starting with a high-strength, self-setting synthetic calcium phosphate cement (*Calcibon*®, Biomet-Merck Biomaterials GmbH, Darmstadt, Germany) [10-13, 74] powder. This robust method uses NaCl crystals as the porogen. Granule preparation was achieved within minutes on an automatic sieve shaker, with desired sizes of sieves stacked on top of one another, under clean room conditions. Upon setting of the cement matrix, porogen crystals were leached out of the granules in water to form pores. These granules are already in clinical use (i.e., Calcibon® Granules) as bone substitute materials for repair and reconstruction of bone defects, in combination with bone marrow aspirate, blood or platelet-rich plasma of the patient.

EXPERIMENTAL PROCEDURE

Granule preparation

Powders of Calcibon® (Biomet-Merck Biomaterials GmbH, Darmstadt) were used as the self-setting CaP cement. The cement powder [75] consisted of a mixture of 62.5 wt% tricalcium phosphate (α-TCP), 26.8 wt% dicalcium phosphate anhydrous (DCPA), 8.9 wt% calcium carbonate ($CaCO_3$) and 1.8 wt% hydroxyapatite (HA). An aqueous solution of Na_2HPO_4 (99.9%, Merck, Darmstadt, 3 wt% solution prepared in water) was used as the initiator liquid. NaCl (99.9%, Merck, Darmstadt) was used as the porogen.

Granule preparation method comprised the steps of a) mixing the Calcibon® cement powder (40.0 g) and NaCl (70.0 g) in a Turbula mixer for 90 minutes, b) wetting the powder with an appropriate volume of a mixture of high-purity ethanol (99.9%, Merck, Darmstadt), and 3.5 wt% Na_2HPO_4 aqueous solution (i.e., 6.7 mL ethanol mixed with a 12 mL aliquot of Na_2HPO_4, c) kneading the wet powder body for 4 minutes to form a cake with a special mixer, d) immediately sieving that wet body over the following 2 minutes in an automatic sieve shaker, with multiple sieves, to *in situ* form granules of desired sizes, e) leaching out the NaCl porogen by soaking the granules in a special sterile bath with circulating water at room temperature from 24 to 48 hours, f.) checking the efficiency of leaching (of NaCl) by measuring the conductivity of the washing water, g) optionally, soaking the granules in a 1 wt% Na_2HPO_4 solution for curing, and h) drying the obtained granules at 37°C, followed by gamma-ray sterilization. Granules of spherical geometry, with sizes varying from 2 to 4 mm, were produced by using the above-outlined method on an industrial scale.

Granule characterization

Granules were characterized by scanning electron microscopy (SEM, JEOL630, Jeol Corp., Tokyo, Japan), energy-dispersive X-ray spectroscopy (EDXS, Thermo-Kevex, San Jose, CA), powder X-ray diffraction (XRD, Cu K_α radiation, D5000, Siemens GmbH, Karlsruhe, Germany), Fourier-transformed infrared spectroscopy (FTIR, Nicolet 550, Thermo-Nicolet, Woburn, MA), water absorption [76], density (Pycnometer, AccuPyc 1330, Micromeritics, Norcross, GA), and compressive strength measurements (Model 4500, Instron Deutschland GmbH, Germany).

Samples were coated with a 50 to 70 nm-thick layer of carbon prior to SEM imaging. EDXS analyses were also performed on such carbon-coated samples. XRD data were gathered over the 2θ range of 10 to 60°, with a step scan rate of 0.02 per minute and preset time of 1 second. 35 mA and 40 kV were the respective power and voltage settings of the X-ray diffractometer during operation. FTIR analyses, on the other hand, were performed after diluting the pulverized (into a fine powder) granule samples in KBr at the sample-to-KBr weight ratio of 1:100, followed by pelletizing in a 1 cm steel die at 25 MPa. Granule densities were measured with a standard gas pycnometer. Compressive strength measurements were performed after filling a 2.1 cm diameter stainless steel die cavity with approximately 0.85 g of granules, followed by gentle tapping of the die to facilitate the even packing of granules. An automated Instron universal testing machine was used to push the upper punch into the die cavity at the crosshead speed of 1 mm/min.

RESULTS AND DISCUSSION

The size distribution of granules obtained in one typical setting-sieving batch (i.e., wet cement paste was set *in situ* during sieving) is given in Table 1. The weight percentages given in Table 1 were the average values obtained from ten separate sieving runs. Size distribution could be altered simply by varying or adjusting the L/P (liquid-to-powder) ratio of the cement body placed onto the sieve shaker.

Table I. Granule size distribution

Granule size (mm)	Weight % (± 3%)
2.8 to 4	15
2 to 2.8	35
1.25 to 2	36
1 to 1.25	4
< 1	10

Figure 1a exhibit the low-magnification SEM image of the obtained granules, and the cubic imprints of the embedded NaCl porogen crystals are shown, which were totally leached out upon washing. In other words, NaCl crystals left behind their footprints. Those square nests formed (after dissolution of these NaCl crystals) in the calcium phosphate cement matrix were like the "replicas" of those crystals, and thus the product could be identified according to its manufacturing process. Macropores, being simply dependent on the crystal size of NaCl used in processing, could be readily varied over the range of 50 to 800 μm. However, for the sample shown in Fig. 1a, pore sizes were from 50 to 500 μm. High-magnification SEM micrograph of the granules given in Figure 1b shows their microporous matrices (i.e., the dense looking areas of Fig. 1a). Interconnected micropores ranged in size from 1 to 4 µm. Since the cement of this study, i.e., *Calcibon®*, is an α-TCP-based cement [11], with the major additives of $CaHPO_4$ and $CaCO_3$, upon setting within 7 to 8 minutes, it formed a web of interlocked, intermingling nanosize platelets (as shown in Fig. 1b) of Ca-deficient apatitic CaP with a Ca/P molar ratio of 1.52±0.1. Granulation process used did not alter or destroy this stoichiometry.

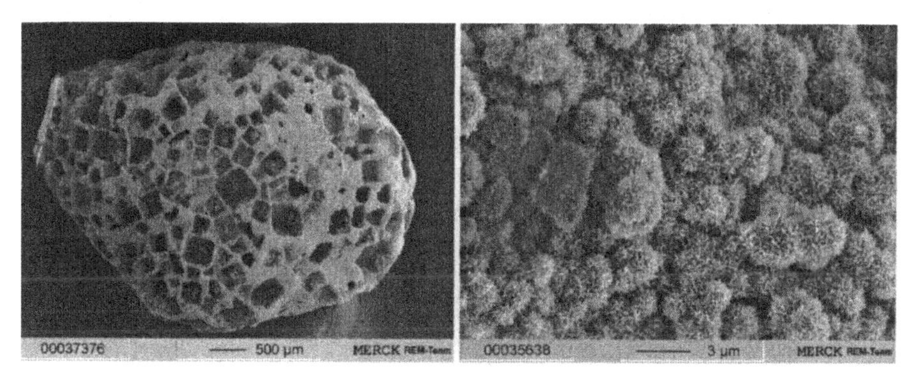

Fig. 1 SEM images of granules; (*left*) macroview, (*right*) microstructure of dense-looking areas

"Biological" or "bone-like" apatite, which constitutes the bone mineral, is known to be a carbonated (4 to 10 wt%), poorly crystallized, alkali (i.e., Na and K) and alkaline earth (Mg) element-doped, non-stoichiometric, calcium-deficient apatitic phosphate with a Ca/P molar ratio variable over the range of 1.50 to 1.70 [1, 77-79]. The CaP cement used in the production of the porous granules of this study was shown [16] to yield a fast deposition of new bone at the cement surface and is considered to be biocompatible [80, 81].

XRD analysis of these granules yielded the characteristic poorly-crystallized apatitic CaP spectra, as shown in Fig. 2a. The setting reaction for this high strength (>55 MPa under compression) cement has been described elsewhere in detail [11]. However, to mention briefly, the major component of this cement, i.e., α-$Ca_3(PO_4)_2$, went through a hydrolysis reaction upon its contact with the setting solution of pH 9. The end product of the hydrolysis reaction was Ca-deficient apatitic CaP. $CaCO_3$ present in the cement powder rapidly participated in this hydrolysis process facilitating the formation of carbonated apatitic CaP [82]. $CaHPO_4$ itself also underwent a similar hydrolysis procedure to the apatitic CaP, and the small amount of precipitated HA present in the cement formulation acted as an accelerator for those hydrolysis processes, which continued at the body temperature till the completion of the cement setting. The XRD data given in Fig. 2a belonged to the freshly produced porous granules after washing off of the NaCl porogen crystals in deionized water at 37°C for 72 hours. Presence of NaCl crystals in granule production by using this cement did not affect the cement setting process.

The carbonated nature of the porous granules was strongly indicated by the characteristic CO_3^{2-} bands seen at 873 and 1450 cm^{-1} in the FTIR data of Fig. 2b. The weak band seen at 3571 cm^{-1} belonged to the OH stretching. Orthophosphate bands were also observed at their characteristic positions [83].

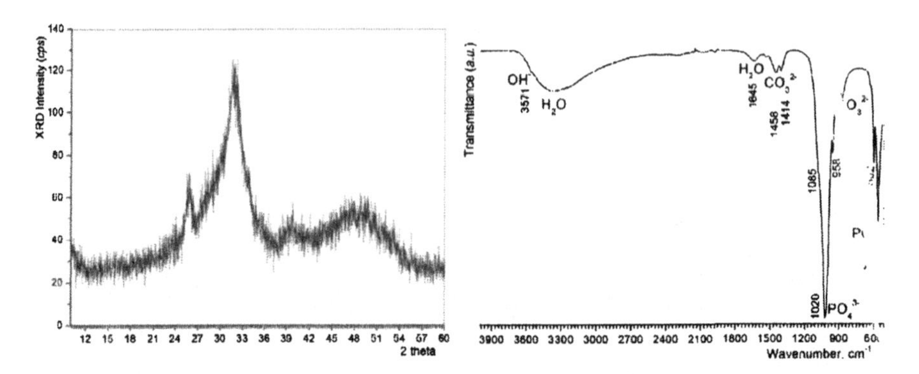

Fig. 2 XRD trace of granules (*left*), FTIR spectrum of granules (*right*)

EDXS analyses performed on the final granules showed that they did not contain any Na or Cl^- ions originating from the use of NaCl as the porogen. EDXS results (accurate to ±0.5 wt%, data not shown), only exhibited the extraneous C peaks due to the carbon coating of granules prior to the analysis.

The formed granules (between 2 to 4 mm sizes) had a pycnometer-measured density of 1.60 ± 0.15 g/cm^3, which corresponded to about 50% total porosity, considering the fact that the density of the fully set, dense *Calcibon®* cement was around 3.2 g/cm^3 [11]. Porous granules also possessed a water absorption percentage of 150, measured according to the aforementioned ASTM standard [76]. This property itself imparted the granules a significant wicking ability. Compressive strengths of the granules were measured to be 10 ± 1 MPa (average of 5 measurements). Porous nature of the granules caused this reduction in compressive strength.

The cement liquid (3.5 wt% Na_2HPO_4 solution), when mixed with the cement powder of this study, resulted in a smooth and malleable calcium phosphate paste. This paste showed the ability of seamlessly embracing and surrounding the cubic NaCl crystals. The use of ethanol in the granule manufacturing was only for the purpose of retarding the setting reaction for a few more minutes to allow successful sieving. Ethanol used was either evaporated or washed away with water during the later, washing and drying stages of the granule manufacturing process. During leaching out of the embedded NaCl crystals, the effluent solutions were monitored with respect to their resistivity on a real time basis. It is known that even ppm levels of dissolved NaCl in water would cause a decrease in the resistivity values of such solutions. Washing operation was terminated when the resistivity of the effluent solutions reached the level of that of pure, distilled water (i.e., 18 MΩ).

The wet, freshly formed granules hardened at the ambient temperature by an endothermic reaction. The chemical composition and crystalline structure of the cured material did mimic the mineral part of natural bone, as depicted in Figures 2(a) and 2(b).

More than a year after the completion of our study on the production of porous CaP granules by the technique described here [73], Tadic *et al.* [84] independently published an article on the manufacture of porous HA objects that avoided sintering. Tadic *et al.* study [84] used a NaCl porogen technique similar to ours to produce macropores with pore diameters in the range of 250 to 400 μm in their 3D objects. Nevertheless, their starting material was just a precipitated hydroxyapatite in powder form, which was far from being a self-setting cement. In order not to cause any undesired grain growth and a decrease in the surface reactivity of their powders, they apparently did not calcine their precipitated powders, and used such powders as received. Tadic *et al.* [84] unfortunately needed to compact their NaCl-embedded hydroxyapatite powder blocks by using cold isostatic pressing. Since they didn't sinter their 3D objects, after porogen removal and consequent pore formation, it was not possible to mention any handling strength to be present in such devices for the orthopedic surgeon. This recent report [84] constituted a vivid example on the significance of using strong self-setting cements in forming porous CaP-based medical devices by the porogen technique.

The granules of the present study regarded to be one-time bone implants staying in place until biodegradation by the body. The *in vivo* resorption rate of porous granules can be theoretically predicted to be much faster in comparison to the dense blocks or prismatic objects made out of the same material [85]. Such granules can be used in every aseptic bone bed, except the area of open epiphyseal discs. Depending on the indication, fixation or stabilization should be performed as if using an autologous bone graft.

CONCLUSIONS

Micro- and macroporous carbonated, apatitic calcium phosphate granules were produced by using the NaCl porogen technique with a self-setting cement powder. The high strength of the CaP cement used as the starting material here facilitated the robust production of porous granules.

(1) The robust process used in granule manufacturing allowed the precise selection and specification of the final granule sizes over the range of 1 to 5 mm,
(2) Pore sizes over the range of 50 to 550 μm were easily achieved,
(3) Granules had about 50% porosity,
(4) Granules were produced from totally synthetic and biocompatible materials, such as α-$Ca_3(PO_4)_2$, $CaHPO_4$, $CaCO_3$ and $Ca_{10}(PO_4)_6(OH)_2$,
(5) Granules comprised carbonated, calcium-deficient, poorly-crystallized apatitic CaP,
(6) Granules had a high wicking ability (*ca.* 150%),
(7) Granule production was performed, from the start to its end, at temperatures between the ambient and 37°C,
(8) Good manufacturing practices (GMP) were implemented at all stages of the granule production.

ACKNOWLEDGEMENTS

A. C. Tas was formerly a staff scientist in "Merck Biomaterials GmbH" and "Biomet-Merck Biomaterials GmbH" in Darmstadt, Germany, from September 2001 to April 2003.

REFERENCES

[1]A. S. Posner and F. Betts, "Synthetic Amorphous Calcium Phosphate and Its Relation to Bone Mineral Structure," *Acc. Chem. Res.*, **8**, 273-81 (1975).

[2]K. D. Rogers and P. Daniels, "An X-ray Diffraction Study of the Effects of Heat Treatment on Bone Mineral Microstructure," *Biomaterials*, **23**, 2577-85 (2002).

[3]A. Bigi, G. Cojazzi, S. Panzavolta, A. Ripamonti, N. Roveri, M. Romanello, K. Suarez, and L. Moro, "Chemical and Structural Characterisation of the Mineral Phase from Cortical and Trabecular Bone," *J. Inorg. Biochem.*, **68**, 45-51 (1997).

[4]J. L. Holden, J. G. Clement, and P. P. Phakey, "Age and Temperature Related Changes to the Ultrastructure and Composition of Human Bone Mineral," *J. Bone Min. Res.*, **10**, 1400-1408 (1995).

[5]E. J. Wheeler and D. Lewis, "An X-ray Study of the Paracrystalline Nature of Bone Apatite," *Calcif. Tissue Res.*, **24**, 243-8 (1977).

[6]R. Handschin and W. Stern, "X-ray Diffraction Studies on the Lattice Perfection of Human Bone Apatite (crista iliaca)," *Bone*, **16**, 355S-363S (1995).

[7]L.L. Hench, "Bioceramics: From Concept to Clinic," *J. Am. Ceram. Soc.*, **74**, 1487-510 (1991).

[8]R. Gunzburg, M. Szpalski, N. Passuti, and M. Aebi, *The Use of Bone Substitutes in Spine Surgery*, pp. 2-11, Springer-Verlag, Berlin, 2002.

[9]G. Daculsi, N. Passuti, S. Martin, C. Deudon, R. Z. LeGeros, and S. Raher, "Macroporous Calcium Phosphate Ceramic for Long Bone Surgery in Humans and Dogs: Clinical and Histological Study, *J. Biomed. Mater. Res.*, **24**, 379-96 (1990).

[10]I. Khairoun, M. G. Boltong, F. C. M. Driessens, and J. A. Planell, "Effect of Calcium Carbonate on the Compliance of an Apatitic Calcium Phosphate Bone Cement," *Biomaterials*, **18**, 1535-9 (1997).

[11]E. Fernandez, F. J. Gil, S. M. Best, M. P. Ginebra, F. C. M. Driessens, and J. A. Planell, "The Cement Setting Reaction in the $CaHPO_4$-α-$Ca_3(PO_4)_2$ System: An X-ray Diffraction Study," *J. Biomed. Mater. Res.*, **42**, 403-406 (1998).

[12]E. Fernandez, F. J. Gil, M. P. Ginebra, S. M. Best, F. C. M. Driessens, and J. A. Planell, "Calcium Phosphate Bone Cements for Clinical Applications, Part I: Solution Chemistry," *J. Mater. Sci. Mater. M.*, **10**, 169-76 (1999).

[13]E. Fernandez, F. J. Gil, M. P. Ginebra, S. M. Best, F. C. M. Driessens, and J. A. Planell, "Production and Characterization of New Calcium Phosphate Bone Cements in the $CaHPO_4$-α-$Ca_3(PO_4)_2$ System: pH, Workability and Setting Times," *J. Mater. Sci. Mater. M.*, **10**, 223-30 (1999).

[14]B. Knepper-Nicolai, A. Reinstorf, I. Hofinger, K. Flade, R. Wenz, and W. Pompe, "Influence of Osteocalcin and Collagen I on the Mechanical and Biological Properties of Biocement D," *Biomolec. Eng.*, **19**, 227-31 (2002).

[15]E. M. Ooms, J. G. C. Wolke, J. P. C. M. van der Waerden, and J. A. Jansen, "Use of Injectable Calcium Phosphate Cement for the Fixation of Titanium Implants: An Experimental Study in Goats," *J. Biomed. Mater. Res.*, **66B**, 447-56 (2003).

[16]P. Q. Ruhe, H. C. Kroese-Deutman, J. G. C. Wolke, P. H. M. Spauwen, and J. A. Jansen, "Bone Inductive Properties of rhBMP-2 loaded Porous Calcium Phosphate Cement Implants in Cranial Defects in Rabbits," *Biomaterials*, **25**, 2123-32 (2004).

[17]K. Kurashina, H. Kurita, M. Hirano, A. Kotani, C. P. A. T. Klein, and K. de Groot, "In Vivo Study of Calcium Phosphate Cements: Implantation of an α-Tricalcium Phosphate/Dicalcium Phosphate Dibasic/Tetracalcium Phosphate Monoxide Cement Paste," *Biomaterials*, **18**, 539-43 (1997).

[18]B. R. Constantz, B. M. Barr, I. C. Ison, M. T. Fulmer, D. C. Delaney, J. Ross, and R. D. Poser, "Histological, Chemical and Crystallographic Analysis of Four Calcium Phosphate Cements in Different Rabbit Osseous Sites," *J. Biomed. Mater. Res.*, **43**, 451-61 (1998).

[19]D. Knaack, M. E. P. Goad, M. Aiolova, C. Rey, A. Tofighi, P. Chakravarthy, and D. D. Lee, "Resorbable Calcium Phosphate Bone Substitute," *J. Biomed. Mater. Res.*, **43**, 399-409 (1998).

[20]C. Knabe, G. Berger, R. Gildenhaar, C. R. Howlett, B. Markovic, and H. Zreiqat, "The Functional Expression of Human Bone-derived Cells Grown on Rapidly Resorbable Calcium Phosphate Ceramics," *Biomaterials*, **25**, 335-44 (2004).

[21]B. L. Eppley, S. Stal, L. Hollier, and M. Kumar, "Compartmentalized Bone Regeneration of Cranial Defects with Biodegradable Barriers - Effects of Calcium Sodium Phosphate Surface Coatings on LactoSorb," *J. Craniof. Surg.*, **13**, 681-6 (2002).

[22]H. Oonishi, L. L. Hench, J. Wilson, F. Sugihara, E. Tsuji, S. Kushitani, and H. Iwaki, "Comparative Bone Growth Behavior in Granules of Bioceramic Materials of Various Sizes," *J. Biomed. Mater. Res.*, **44**, 31-43 (1999).

[23]H. Gauthier, G. Daculsi, and C. Merle, "Association of Vancomycin and Calcium Phosphate by Dynamic Compaction: In Vitro Characterization and Microbiological Activity," *Biomaterials*, **22**, 2481-7 (2001).

[24]K. A. Hing, S. M. Best, K. E. Tanner, P. A. Revell, and W. Bonfield, "Quantification of Bone Ingrowth within Bone-derived Porous Hydroxyapatite Implants of Varying Density," *J. Mater. Sci. Mater. M.*, **10**, 663-70 (1999).

[25]F. H. Lin, C. J. Liao, K. S. Chen, and J. S. Sun, "Preparation of a Biphasic Porous Bioceramic by Heating Bovine Cancellous Bone with $Na_4P_2O_7 \cdot 10H_2O$ Addition," *Biomaterials*, **20**, 475-84 (1999).

[26]S. Joschek, B. Nies, R. Krotz, and A. Gopferich, "Chemical and Physicochemical Characterization of Porous Hydroxyapatite Ceramics Made of Natural Bone," *Biomaterials*, **21**, 1645-58 (2000).

[27]R. Bareille, M. H. Lafage-Proust, C. Faucheux, N. Laroche, R. Wenz, M. Dard, and J. Amedee, "Various Evaluation Techniques of Newly Formed Bone in Porous Hydroxyapatite Loaded with Human Bone Marrow Cells Implanted in an Extra-Osseous Site," *Biomaterials*, **21**, 1345-52 (2000).

[28]P. Weiss, L. Obadia, D. Magne, X. Bourges, C. Rau, T. Weitkamp, I. Khairoun, J.M. Bouler, D. Chappard, O. Gauthier, and G. Daculsi, "Synchrotron X-ray Microtomography (on a micron scale) Provides Three-dimensional Imaging Representation of Bone Ingrowth in Calcium Phosphate Biomaterials," *Biomaterials*, **24**, 4591-601 (2003).

[29]N. O. Engin and A. C. Tas, "Preparation of Porous $Ca_{10}(PO_4)_6(OH)_2$ and β-$Ca_3(PO_4)_2$ Bioceramics," *J. Am. Ceram. Soc.*, **83**, 1581-4 (2000).

[30]H. S. Cheung, "In vitro Cartilage Formation on Porous Hydroxyapatite Ceramic Granules," *In Vitro Cell. Dev. B.*, **21**, 353-7 (1985).

[31]K. Ono, T. Yamamuro, T. Nakamura, and T. Kokubo, "Quantitative Study on Osteoconduction of Apatite Wollastonite-containing Glass Ceramic Granules, Hydroxyapatite Granules, and Alumina Granules," *Biomaterials*, **11**, 265-71 (1990).

[32]A. M. Gatti, D. Zaffe, and G. P. Poli, "Behavior of Tricalcium Phosphate and Hydroxyapatite Granules in Sheep Bone Defects," *Biomaterials*, **11**, 513-7 (1990).

[33]J. M. Sautier, J. R. Nefussi, H. Boulekbache, and N. Forest, "In vitro Bone Formation on Coral Granules," *In Vitro Cell. Dev. B.*, **26**, 1079-85 (1990).

[34]K. Yamamura, H. Iwata, and T. Yotsuyanagi, "Synthesis of Antibiotic-loaded Hydroxyapatite Beads and Drug Release Testing," *J. Biomed. Mater. Res.*, **26**, 1053-64 (1992).

[35]H. Yamasaki and H. Sakai, "Osteogenic Response to Porous Hydroxyapatite Ceramics under the Skin of Dogs," *Biomaterials*, **13**, 308-12 (1992).

[36]H. S. Byrd, P. C. Hobar, and K. Shewmake, "Augmentation of the Craniofacial Skeleton with Porous Hydroxyapatite Granules," *Plast. Reconstr. Surg.*, **91**, 15-22 (1993).

[37]K. G. Wiese and H. A. Merten, "The Role of the Periosteum in Osteointegration of Hydroxyapatite Granules," *Int. J. Oral Max. Surg.*, **22**, 306-8 (1993).

[38]T. C. Lindholm, T. J. Gao, and T. S. Lindholm, "The Role of Autogeneic Bone-marrow in the Repair of a Skull Trephine Defect filled with Hydroxyapatite Granules in the Rabbit," Int. *J. Oral Max. Surg.*, **23**, 306-11 (1994).

[39]M. Fabbri, G. C. Celotti, and A. Ravaglioli, "Granulates Based on Calcium Phosphate with Controlled Morphology and Porosity for Medical Applications: Physico-chemical Parameters and Production Technique," *Biomaterials*, **15**, 474-7 (1994).

[40]A. W. Sugar, P. Thielens, G. D. Stafford, and M. J. Willins, "Augmentation of the Atrophic Maxillary Alveolar Ridge with Hydroxyapatite Granules in a Vicryl (Polyglactin-910) Knitted Tube and Simultaneous Open Vestibuloplasty," *Brit. J. Oral Max. Surg.*, **33**, 93-7 (1995).

[41]E. A. Holtgrave and K. Donath, "Response of Odontoblast-like Cells to Hydroxyapatite Ceramic Granules," *Biomaterials*, **16**, 155-9 (1995).

[42]P. Luo and T. G. Nieh, "Preparing Hydroxyapatite Powders with Controlled Morphology," *Biomaterials*, **17**, 1959-64 (1996).

[43]S. Takahashi and Y. Nakano, "A Morphological Study on Obliteration of the Temporal Dorsal Bullae using Hydroxyapatite Granules," *Am. J. Otol.*, **17**, 197-9 (1996).

[44]D.-M. Liu, "Fabrication and Characterization of Porous Hydroxyapatite Granules," *Biomaterials*, **17**, 1955-7 (1996).

[45]A. E. Ingram, J. Robinson, and R. J. Rohrich, "The Antibacterial Effect of Porous Hydroxyapatite Granules," *Plast. Reconstr. Surg.*, **98**, 1119 (1996).

[46]M. Maruyama and M. Ito, "In Vitro Properties of a Chitosan-bonded Self-hardening Paste with Hydroxyapatite Granules," *J. Biomed. Mater. Res.*, **32**, 527-32 (1996).

[47]N. Kawai, S. Niwa, M. Sato, Y. Sato, Y. Suwa, and I. Ichihara, "Bone Formation by Cells from Femurs Cultured among Three-dimensionally Arranged Hydroxyapatite Granules," *J. Biomed. Mater. Res.*, **37**, 1-8 (1997).

[48]M. Vasconcelos, A. Afonso, R. Branco, and J. Cavalheiro, "Guided Bone Regeneration using Osteoapatite Granules and Polytetrafluoroethylene Membranes," *J. Mater. Sci. Mater. M.*, **8**, 815-8 (1997).

[49]V. Mooney, J. B. Massie, B. I. Lind, J.-H. Rah, S. Negri, and R. E. Holmes, "Comparison of Hydroxyapatite Granules to Autogeneous Bone Graft in Fusion Cages in a Goat Model," *Surg. Neurol.*, **49**, 623-33 (1998).

[50]H. Gauthier, J. Guicheux, G. Grimandi, A. Faivre-Chauvet, G. Daculsi, and C. Merle, "In Vitro Influence of Apatite-Granule-Specific Area on Human Growth Hormone Loading and Release," *J. Biomed. Mater. Res.*, **40**, 606-13 (1998).

[51]M. A. W. Merkx, J. C. Maltha, H. P. M. Freihofer, and A. M. Kuijpers-Jagtman, "Incorporation of Particulated Bone Implants in the Facial Skeleton," *Biomaterials*, **20**, 2029-35 (1999).

[52]O. Malard, J. M. Bouler, J. Guicheux, D. Heymann, P. Pilet, C. Coquard, and G. Daculsi, "Influence of Biphasic Calcium Phosphate Granulometry on Bone Ingrowth, Ceramic Resorption, and Inflammatory Reactions: Preliminary in vitro and in vivo Study," *J. Biomed. Mater. Res.*, **46**, 103-11 (1999).

[53]N. Ikeda, K. Kawanabe, and T. Nakamura, "Quantitative Comparison of Osteoconduction of Porous, Dense A-W Glass-ceramic and Hydroxyapatite Granules (effects of granule and pore sizes)," *Biomaterials*, **20**, 1087-95 (1999).

[54]W. Paul and C. P. Sharma, "Development of Porous Spherical Hydroxyapatite Granules: Application towards Protein Delivery," *J. Mater. Sci. Mater. M.*, **10**, 383-8 (1999).

[55]C. Schwartz, P. Liss, B. Jacquemaire, P. Lecestre, and P. Frayssinet, "Biphasic Synthetic Bone Substitute Use in Orthopedic and Trauma Surgery: Clinical, Radiological and Histological Results," *J. Mater. Sci. Mater. M.*, **10**, 821-5 (1999).

[56]H. Oonishi, Y. Kadoya, H. Iwaki, and N. Kin, "Hydroxyapatite Granules Interposed at Bone-Cement Interface in Total Hip Replacements: Histological Study of Retrieved Specimens," *J. Biomed. Mater. Res.*, **53**, 174-80 (2000).

[57]H. Yano, H. Ohashi, Y. Kadoya, A. Kobayashi, Y. Yamano, and Y. Tanabe, "Histologic and Mechanical Evaluation of Impacted Morcellized Cancellous Allografts in Rabbits - Comparison with Hydroxyapatite Granules," *J. Arthroplasty*, **15**, 635-43 (2000).

[58]P. C. Hobar, M. Pantaloni, and H. S. Byrd, "Porous Hydroxyapatite Granules for Alloplastic Enhancement of the Facial Region," *Clin. Plast. Surg.*, **27**, 557 (2000).

[59]A. M. Gatti, E. Monari, G. Poli, and E. Galli, "Clinical Long-term Evaluation of Hydroxyapatite Granules Implantation in Periodontal Defects," *Key Eng. Mater.*, **192**, 921-4 (2000).

[60]H. G. Baramki, T. Steffen, P. Lander, M. Chang, and D. Marchesi, "The Efficacy of Interconnected Porous Hydroxyapatite in Achieving Posterolateral Lumbar Fusion in Sheep," *Spine*, **25**, 1053-60 (2000).

[61]S. Takagi and L. C. Chow, "Formation of Macropores in Calcium Phosphate Cement Implants, " *J. Mater. Sci. Mater. M.*, **12**, 135-9 (2001).

[62]A. John, S. Abiraman, H. K. Varma, T. V. Kumari, and P. R. Umashankar, "Bone Growth Response with Porous Hydroxyapatite Granules in a Critical Sized Lapine Tibial-defect Model," *Bull. Mater. Sci.*, **25**, 141-54 (2002).

[63]V. S. Komlev, S. M. Barinov, and E. V. Koplik, "A Method to Fabricate Porous Spherical Hydroxyapatite Granules Intended for Time-controlled Release," *Biomaterials*, **23**, 3449-54 (2002).

[64]V. S. Komlev and S. M. Barinov, "Porous Hydroxyapatite Ceramics of Bi-modal Pore Size Distribution," *J. Mater. Sci. Mater. M.*, **13**, 295-9 (2002).

[65]E. Krylova, A. Ivanov, V. Orlovski, G. El-Registan, and S. M. Barinov, "Hydroxyapatite-Polysaccharide Granules for Drug Delivery," *J. Mater. Sci. Mater. M.*, **13**, 87-90 (2002).

[66]N. Patel, S. M. Best, W. Bonfield, I. R. Gibson, K. A. Hing, E. Damien, and P. A. Revell, "A Comparative Study on the In Vivo Behavior of Hydroxyapatite and Silicon Substituted Hydroxyapatite Granules," *J. Mater. Sci. Mater. M.*, **13**, 1199-1206 (2002).

[67]L. M. Rodriguez-Lorenzo, M. Vallet-Regi, and J. M. F. Ferreira, "Fabrication of Porous Hydroxyapatite Bodies by a New Direct Consolidation Method: Starch Consolidation," *J. Biomed. Mater. Res.*, **60**, 232-40 (2002).

[68]A. Moreira-Gonzalez, I. T. Jackson, T. Miyawaki, V. DiNick, and R. Yavuzer, "Augmentation of the Craniomaxillofacial Region using Porous Hydroxyapatite Granules," *Plast. Reconstr. Surg.*, **111**, 1808-17 (2003).

[69]C. Tanaka, J. Shikata, M. Ikenaga, and M. Takahashi, "Acetabular Reconstruction using a Kerboull-type Acetabular Reinforcement Device and Hydroxyapatite Granules - A 3-to 8-year Follow-up Study," *J. Arthroplasty*, **18**, 719-25 (2003).

[70]A. Sari, R. Yavuzer, S. Ayhan, S. Tuncer, O. Latifoglu, K. Atabay and M. C. Celebi, "Hard Tissue Augmentation of the Mandibular Region with Hydroxyapatite Granules," *J. Craniofac. Surg.*, **14**, 919-23 (2003).

[71]E. T. Baran, K. Tuzlakoglu, A. J. Salgado, and R. L. Reis, "Multichannel Mould Processing of 3D Structures from Microporous Coralline Hydroxyapatite Granules and Chitosan Support Materials for Guided Tissue Regeneration/Engineering," *J. Mater. Sci. Mater. M.*, **15**, 161-5 (2004).

[72]Z. Zyman, V. Glushko, V. Filippenko, V. Radchenko, and V. Mezentsev, "Nonstoichiometric Hydroxyapatite Granules for Orthopaedic Applications," *J. Mater. Sci. Mater. M.*, **15**, 551-8 (2004).

[73]A. C. Tas, "Method of Preparing Porous Calcium Phosphate Granules," Patent Appl. No^s: WO 03093196 (Publication Date: November 13, 2003, Priority Date: May 06, 2002), US2005/0260115, EP 1501771, AU 2003/229615, CA 2483859.

[74]R. P. del Real, E. M. Ooms, J. G. C. Wolke, M. Vallet-Regi, and J. A. Jansen, "In Vivo Bone Response to Porous Calcium Phosphate Cement," *J. Biomed. Mater. Res.*, **65**, 30-6 (2003).

[75]R. Wenz and F. C. M. Driessens, "Biocements having Improved Compressive Strength," US Patent No: 6,495,156. December 17, 2002.

[76]"Standard Test Method for Apparent Porosity, Water Absorption, Apparent Specific Gravity and Bulk Density of Burned Refractory Brick and Shapes by Boiling Water," ASTM Designation C20-92. 1995 Annual Book of ASTM Standards, Vol. 15.01; pp. 5–7. American Society for Testing and Materials, Philadelphia, PA.

[77]E. J. Wheeler and D. Lewis, "An X-ray study of the paracrystalline nature of bone apatite," *Calcif. Tissue Res.*, **24**, 243-8 (1977).

[78]J. L. Holden, J. G. Clement, and P. P. Phakey, "Age and temperature related changes to the ultrastructure and composition of human bone mineral," *J. Bone Min. Res.*, **10**, 1400-1408 (1995).

[79]A. Bigi, G. Cojazzi, S. Panzavolta, A. Ripamonti, N. Roveri, M. Romanello, K. Suarez, and L. Moro, "Chemical and structural characterisation of the mineral phase from cortical and trabecular bone," *J. Inorg. Biochem.*, **68**, 45-51 (1997).

[80]E. M. Ooms, J. G. C. Wolke, J. P. van der Waerden, and J. A. Jansen, "Trabecular Bone Response to Injectable Calcium Phosphate (Ca-P) Cement," *J. Biomed. Mater. Res.*, **61**, 9–18 (2002).

[81]E. M. Ooms, J. G. C. Wolke, M. T. van de Heuvel, B. Jeschke, and J. A. Jansen, "Histological evaluation of the bone response to calcium phosphate cement implanted in cortical bone," *Biomaterials*, **24**, 989-1000 (2003).

[82]M. P. Ginebra, E. Fernandez, F. C. M. Driessens, and J. A. Planell, "Modeling of the Hydrolysis of α-Tricalcium Phosphate," *J. Am. Ceram. Soc.*, **82**, 2808-12 (1999).

[83]A. C. Tas, "Synthesis of Biomimetic Ca-Hydroxyapatite Powders at 37°C in Synthetic Body Fluids," *Biomaterials*, **21**, 1429-38 (2000).

[84]D. Tadic, F. Beckmann, K. Schwarz, and M. Epple, "A Novel Method to Produce Hydroxyapatite Objects with Interconnecting Porosity that Avoids Sintering," *Biomaterials*, **25**, 3335-40 (2004).

[85]M. Bohner and F. Baumgart, "Theoretical Model to Determine the Effects of Geometrical Factors on the Resorption of Calcium Phosphate Bone Substitutes," *Biomaterials*, **25**, 3569-82 (2004).

A SELF-SETTING, MONETITE ($CaHPO_4$) CEMENT FOR SKELETAL REPAIR

Tarang R. Desai, Sarit B. Bhaduri, and A. Cuneyt Tas
School of Materials Science and Engineering
Clemson University
Clemson, SC 29634

ABSTRACT

A low compressive strength (2 to 4 MPa) but very simple and inexpensive orthopedic $CaHPO_4$ cement was developed by using commercially available $Ca(OH)_2$ powders as the only starting powder component. The setting solution used was a special aqueous phosphoric acid (H_3PO_4) solution, with a small amount of sodium hydrogen carbonate ($NaHCO_3$) dissolved in it. Calcium was provided by the powder component, and all of the phosphate came from the setting solution. The setting solution was acidic, and this helped to neutralize $Ca(OH)_2$ only to the extent of forming the phase of dicalcium phosphate anhydrous (DCPA, monetite), $CaHPO_4$. At an L/P (liquid-to-powder) ratio of 1.54 and Ca/P molar ratio of 0.9, the cements had an initial setting time of 19±2 minutes. Set cements comprised only crystalline $CaHPO_4$. Monetite has a higher solubility than octacalcium phosphate, β-tricalcium phosphate and calcium hydroxyapatite in aqueous solutions at the physiological pH. This new cement was considered to exhibit a higher *in vivo* resorbability in comparison to the apatitic cements. Cement samples were characterized by XRD, FTIR and SEM. Compressive strength, initial and final setting times (with the use of Gillmore needles) of the cement samples were also reported.

INTRODUCTION

Beyond the autologous bone grafts, synthetic calcium phosphates (CaP) are the materials of choice for skeletal repair, mainly due to their biocompatibility, bioactivity, osteoconductivity, and even bioresorbability (i.e., osteoclast cell-mediated resorption). CaP based cements were first commercialized about two decades ago[1,2]. Self-setting or self-hardening (upon reacting the supplied CaP-based powder component with a special setting or starter solution) CaP cements provided the orthopedic surgeon with a valuable tool of filling three dimensional bone defect within minutes in the surgical theater. Until that time preshaped hard blocks or pieces of CaP were needed to be chiseled and hammered down by the surgeon to the suitable shape and size in accordance with the defect peculiarity.

Most CaP-based orthopedic cements available today were designed to set into calcium hydroxyapatite [HA, $Ca_{10}(PO_4)_6(OH)_2$] as their end products, mainly because of the so-called similarity of HA with the mineral portion of human bones. Hydroxyapatite has the lowest solubility among all the calcium phosphate phases.[1] Moreover, as a result of recent *in vivo* tests,[3,4] such cements based on HA were shown to suffer from low *in vivo* resorbability, and lacking full participation in the bone remodeling processes even a year after their implantation. HA-based cements typically used α-tricalcium phosphate [α-TCP, α-$Ca_3(PO_4)_2$] or tetracalcium phosphate [TTCP, $Ca_4(PO_4)_2O$] as their major constituents. These phases both are able to undergo hydrolysis, into apatitic calcium phosphates, when brought into contact with aqueous solutions. Nevertheless, α-TCP (or TTCP) powders cannot be synthesized at room temperature, and both are fully stable only at high temperatures (in excess of 1200°C), therefore, these high-temperature phases must be quenched to the ambient temperature, by using quite capital-intensive manufacturing investments.[5-7] The involvement of high temperatures in synthesizing α-TCP and TTCP powders also places a burden on the particle size of these important cement ingredients. Typically, quenched chunks or hard agglomerates of those materials require a delicate grinding operation, which may be a further source of contamination, or premature and undesired interaction of the powders with humidity (to undergo through a smaller extent of particle surface-limited hydrolysis) present during grinding. As a result, the reactivity of these powders would also be a major point of concern[8].

Currently, CaP cements are only available in two categories; (i) apatitic[1,2,9] and (ii) brushitic[10-12] [dicalcium phosphate dihydrate, DCPD, $CaHPO_4{\cdot}2H_2O$], based on the final phase of the set cements. Recent interest in the development of CaP cements is shifting more towards brushitic cements due to their proven high bioresorbability[13]. The bioresorption rate of brushitic cements, i.e., their capability to take part in the *in vivo* bone remodeling processes, was found to be significantly higher than that of apatitic cements[14].

However, all of the currently known cement formulations were based on a powder component which consists of an intricate mixture of at least two or more CaP or Ca- or P-containing phases[15], usually some of those ingredients are basic and the remaining are acidic in nature. Upon achieving the first contact with the setting solution, neutralization reactions start to take place in the powder components to form the apatite- or brushite-like end phases. Such a mixture of CaP may undergo a solid state reaction, limiting the shelf life of these cements and also their storage conditions[16].

$CaHPO_4$ (monetite) was reported[17,18] to have a very high aqueous solubility at and around physiological pH values compared to all other technologically important and biocompatible CaP phases, such as octacalcium phosphate [$Ca_8H_2(PO_4)_6{\cdot}5H_2O$], β-tricalcium phosphate and calcium hydroxyapatite. This high solubility has been the main stimulus for the current study, which aimed to prove the feasibility of forming the very first orthopedic cement based on $CaHPO_4$. Monetite was previously studied for applications in the treatment of dental caries and lesions as a constituent of chewing gums[19]. We have observed that (data not reported here) upon its immersion in a Synthetic Body Fluid (SBF) solution[20], $CaHPO_4$ converted into poorly-crystallized, carbonated apatitic calcium phosphate in a relatively short time (less than 48 hours at 37°C), which can be interpreted as an indicator of the highly promising bioactivity of this material. *In vitro* or *in vivo* biocompatibility of $CaHPO_4$ has not yet been reported, and we will soon publish the very first osteoblast cell culture (*in vitro*) results for this bioactive bone substitute material, in direct comparison to brushite, β-tricalcium phosphate, and hydroxyapatite.

To the best of our knowledge, a CaP cement which set into single-phase $CaHPO_4$ has not been reported yet. Such a $CaHPO_4$ cement scaffold was regarded to exhibit increased resorbability, both *in vitro* and *in vivo*, when compared to apatitic CaP cements.[21] Moreover, such inexpensive cement scaffolds may find a wider range of applications (beyond orthopedic uses) in the future as a class of "perfectly non-toxic and biocompatible" materials with easier formability. Our aim here was to synthesize monetite in cement form using a single, inexpensive, off-the-shelf powder component, i.e., calcium hydroxide ($Ca(OH)_2$), and phosphoric acid (H_3PO_4) as the major component of the setting solution.

EXPERIMENTAL PROCEDURE

Starting powder and setting solution

The powder component (*P*) of the cement consisted only of $Ca(OH)_2$ (>95%, Fisher Scientific, Fair Lawn, NJ), and it was used as-received, without any further treatments or additives. Two setting solutions (*SS*) were experimented with; setting solution-1 (*SS-1*) consisted of pure orthophosphoric acid (86.2%, H_3PO_4, Fisher Scientific, Fair Lawn, NJ). Setting solution-2 (*SS-2*) was prepared by mixing H_3PO_4, sodium hydrogen carbonate (>99.7%, $NaHCO_3$, Acros-Fisher Chemicals, Fair Lawn, NJ), and de-ionized water (DI, Millipore, 18.2 MW) in a glass beaker. For a typical preparation of SS-2, 1 g of $NaHCO_3$ was first placed in a 50 mL-capacity beaker and 3 mL of DI was added. This mixture was stirred with a glass rod for 2 minutes, followed by the slow addition of 12 mL of H_3PO_4 through a pipette or burette. Therefore, one was able to prepare 15 mL of the SS-2 at a time. Stable setting solutions were then stored at room temperature in tightly-capped glass media bottles.

Cement synthesis

For the cement synthesis, the setting solution *(SS-1 or SS-2)* was taken in a 10 mL-capacity glass vial and different amounts of DI were added to it to make *SS* with different volumetric proportions of DI. By this way, the amount of liquid (*L*) could be manipulated without changing the number of moles of P in the setting solution. The *SS + DI* (*=L*) solution was manually shaken in that small glass vial for 30 seconds. This liquid, *L*, was then added at once to the accurately weighed powder (*P*), i.e., $Ca(OH)_2$, which was previously placed into an agate mortar. The mixture was manually kneaded by using an agate pestle for a time at the end of which a paste-like substance could be easily scraped out of the mortar with the finger tips. By using different amounts of DI with the setting solution, the L/P ratio was easily changed. However, to obtain a consistent paste with good handling properties, 1.235 g of $Ca(OH)_2$ was placed in the mortar, 1.5 mL of *SS* was taken in the glass vial to which 0.4 mL of DI was then added. This optimized recipe corresponded to an L/P ratio of 1.54. The liquid was shaken for 30 seconds and then added to the powder, which after mixing for approximately 3 minutes yielded a good paste.

Characterization

Characterization of the starting powder i.e. $Ca(OH)_2$ included Fourier Transformed Infrared Spectroscopy (FT-IR, Nicolet 550, Thermo-Nicolet, Woburn, MA), scanning electron microscopy (SEM, FE-SEM, S-4700 and S-4800, Hitachi, Tokyo, Japan), and inductively coupled plasma atomic emission spectroscopy (ICP-AES, Model 61E, Thermo Jarrell Ash, Thermo Electron, Madison WI) analysis. Phase identification in set cement powder samples was perfomed by using powder X-ray diffraction (XDS-2000, Scintag, Sunnyvale, CA), operated at 40 kV and 30 mA, equipped with a Cu K_α-tube at a step size of 0.02° 2θ. FT-IR analyses were performed by Attenuated Total Reflection (ATR) method on powdered cement samples, using a diamond ATR window. Morphology of the cement samples was studied using SEM. Prior to the SEM studies, samples were coated with a thin layer of Pt to improve their conductivity. ICP-AES analysis was also used for the determination of Ca/P molar ratios in the set cement samples.

Compressive strength and setting time measurements

Cylindrical samples with an aspect ratio (length/diameter) of about 1.5 were prepared using a steel die. The formed cylinders were 18 ± 0.05 mm long, with a diameter of 12 ± 0.02 mm. To mention the procedure briefly, the paste was mixed in the agate mortar for a time that was required to get a good consistency depending on the L/P ratio studied. The paste was then manually forced into the cavity of the steel die, and a metal block with a weight corresponding to a pressure of only 0.775 MPa was placed on top of the mold for a period of 15 minutes. After 15 minutes, cylinders obtained were stored at 37±1°C for 36 h. Post incubation period, cylinder dimensions were measured using a digital caliper and compressive strength (CS) tests were carried out on a SATEC-Apex Universal testing machine by using a 4,500 kg load cell. A crosshead speed of 0.5 mm/min was used during the compressive loading runs.

A Gillmore needle apparatus was built and then used to measure the setting time of the cement. A weight of 453.6 g was used for the heavier stainless steel needle and 115.12 g for the lighter needle. t_I was the time in minutes when the lighter needle did not leave an indentation deeper than 1 mm on the cement surface and t_F was the time in minutes when the heavier needle failed to leave an indentation deeper than 1 mm on the cement surface.

RESULTS & DISCUSSION

Figure 1 showed the SEM pictures of the starting $Ca(OH)_2$ powders. As seen from these pictures, $Ca(OH)_2$ particles (particle size ranging from 2 to 10 μm) have a layered or stacked sheets-like microstructure. ICP analyses of the $Ca(OH)_2$ powders showed that the powders also contained 2600±150 ppm Mg. Magnesium is known to be an inhibitor of apatite crystallization,[22] and this proved to be an advantage about the selection of these powders for $CaHPO_4$-cement synthesis. Figure 2a depicted the IR spectra of the $Ca(OH)_2$ powders and it can be seen that there was some conversion of $Ca(OH)_2$ to $CaCO_3$ during storage at room temperature.

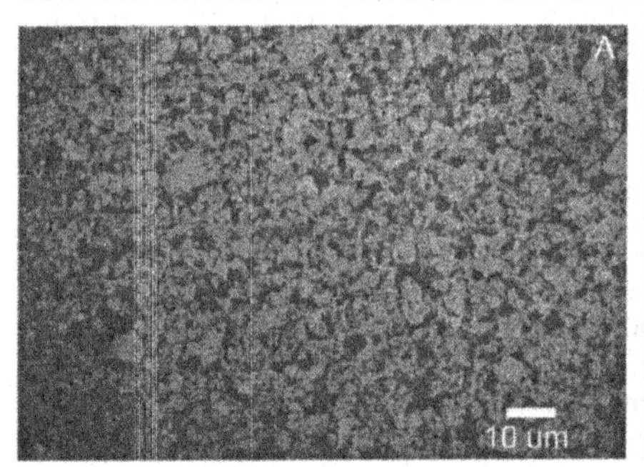

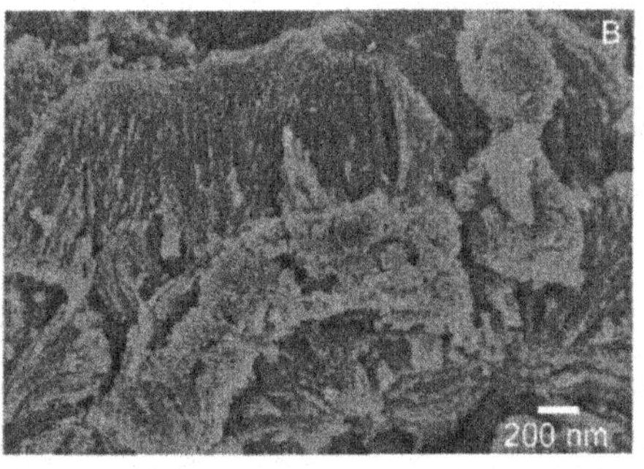

Fig. 1 (*a*) low magnification SEM image of the starting $Ca(OH)_2$ powders, and (*b*) a higher magnification image of the same

Preliminary experiments conducted by using SS-1, i.e., $H_3PO_4 + H_2O$ mixtures, led to the presence of monocalcium phosphate monohydrate (MCPM, $Ca(H_2PO_4)_2 \cdot H_2O$) in all the samples. This was not surprising at all, high acidity in the cement pastes always led to the formation of more acidic calcium phosphates. The two IR spectra given in Figure 2b are obtained from samples prepared by using separate Ca/P molar ratios of 0.5 (top trace) and 0.9 (bottom trace).

An L/P ratio of 1.86 was used in both cases and the sample with a Ca/P molar ratio of 0.5 showed pure MCPM, while Ca/P molar ratio of 0.9 resulted in a more or less biphasic mixture of $CaHPO_4$ and MCPM. SEM analysis of these two samples also showed the characteristic large plate-like crystals of MCPM for the sample with Ca/P molar ratio 0.5 (shown in Figure 2c). The SEM image of sample with a Ca/P ratio of 0.9 showed two distinct phases, one growing as globules on top of the other, which is also confirmed by the FTIR spectra. Since the setting solution contained only H_3PO_4, the pH value of the system went down to a very low value, hence the observation of MCPM was expected.[23] Moreover, since MCPM had a very high rate of nucleation under the specific synthesis conditions we used, the materials obtained did not show any extent of self-setting cement characteristics and turned into a hard mass very rapidly in the mortar.

The amount of MCPM obtained in the final mixtures decreased in going from the Ca/P molar ratio of 0.5 to that of 0.9, and since the theoretical Ca/P molar ratio of the aimed stoichiometry (monetite, $CaHPO_4$) was 1.0, further experiments were performed with a Ca/P ratio of 0.9. Effect of different L/P ratios on the final phase composition was also studied using SS-1, and MCPM was found to be present in all samples at all L/P ratios although the amount of MCPM varied; a higher L/P ratio led to decreased MCPM amounts in the final material. It can be hypothesized that the fluid pastes obtained by using higher L/P ratios would have higher pH values compared to those obtained by using comparatively lower L/P ratios. This slightly higher pH combined with better mixing due to increased amounts of DI could be one reason for the observation of decreasing amounts of MCPM at higher L/P ratios.

MCPM is an extremely acidic phase and for that reason it can neither be desirable nor used alone as a biomaterial for skeletal repair. Highly acidic calcium phosphates, such as MCPM, cause immediate inflammatory response upon implantation. On the other hand, the inflammatory response to brushitic cements was found to be quite mild and transient in nature.[13] Since the very low pH of the setting solution was the sole reason for MCPM formation, an additive which would increase the pH of the setting solution only slightly, such as $NaHCO_3$, was regarded to alleviate the problem of MCPM formation in the cement. Different amounts of $NaHCO_3$ in a new setting solution (i.e., SS-2) were experimented with, and a very small amount of $NaHCO_3$ addition (0.066 g/mL) to the setting solution (as described in the Experimental section) was found to be extremely effective in completely eliminating the MCPM phase from the final set cements as seen from the IR and XRD data given in Figures 3a and 3b, respectively.

The development of the new SS-2, therefore, resulted in obtaining pure $CaHPO_4$ as the final product of the set cements. ICP analyses of the set cements gave the Ca and P wt% values of 24.91 and 19.86, respectively. This translated to a Ca/P molar ratio of 0.96 in the solid cement samples, which was in close agreement with the theoretical Ca/P molar ratio for $CaHPO_4$ (1.0). With a further increase in the concentration of $NaHCO_3$ in the setting solution (i.e., SS-2 derivatives), apatitic CaP formation was seen along with monetite (results not shown). Some unreacted $Ca(OH)_2$ was also observed in the XRD plot, which did not readily show up in the IR spectra.

SEM photomicrographs of Figures 3c through 3e showed the microstructure of the set cement samples. At a low magnification (Fig. 3c), the cement was seen to contain a significant amount of pores which were interconnected (observed at higher magnifications). Porous cements are favorably required to allow for the wicking of blood and blood cells towards the bulk of the cement samples. At higher magnifications (Figs. 3d and 3e), the SEM images also showed that the cement microstructure was quite different from that of the starting powder, i.e., $Ca(OH)_2$.

Interlocking of $CaHPO_4$ plates was also seen at numerous places which imparted the cement its limited but still significant strength.

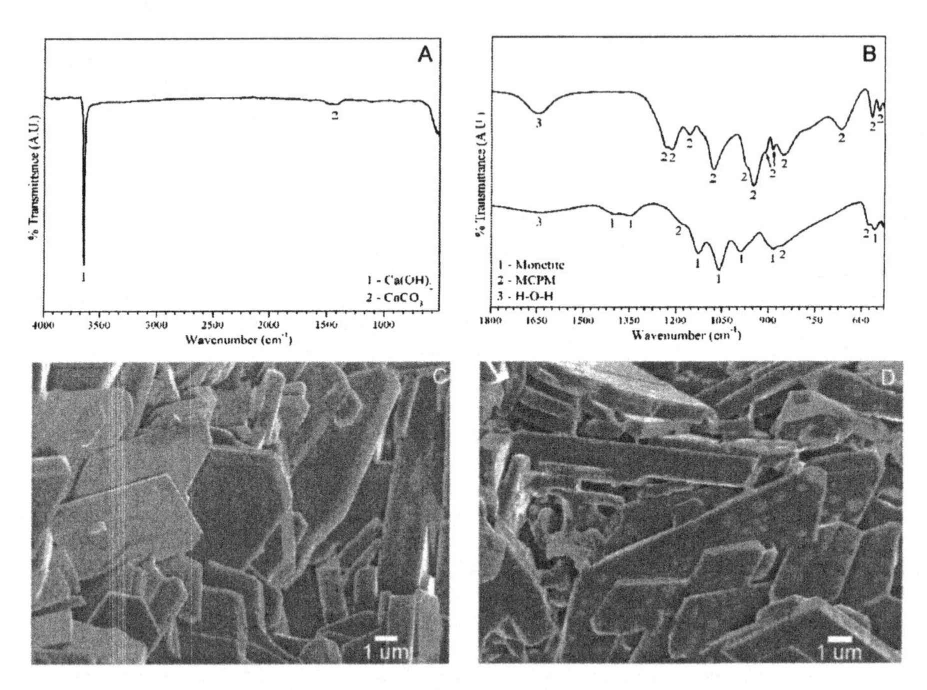

Fig. 2 (*a*) IR spectrum of starting Ca(OH)2 powders, (*b*) bottom: IR spectrum of Ca/P=0.5, SS-1 substance, and top: IR data for Ca/P=0.9, SS-1 substance, (*c*) SEM image of sample obtained with a Ca/P ratio of 0.5, SS-1, MCPM crystals, (*d*) SEM image of a Ca/P ratio of 0.9, SS-1 substance, MCPM crystals together with a small amount of $CaHPO_4$.

The mechanical strength of any orthopedic cement is an important parameter since the cement should have some load-bearing ability during its lifetime as an implant. Figure 3f showed a characteristic compressive strength (CS) chart for the $CaHPO_4$ cement of this study. The CS of the cement was found to be 2.05 ± 0.2 MPa. It should be noted that the human trabecular bones have a modest compressive strength over the range of 2 to 10 MPa.[1,2] The setting time of the cement of this study was determined by using the Gillmore needle apparatus and the optimized cement recipe had an initial setting time, i.e., t_I, of 19 ± 2 minutes and a final setting time, t_F, of 58 ± 3 at an L/P ratio of 1.54.

The rate of nucleation of monetite (with SS-2) was somewhat slower in comparison to the highly acidic MCPM (with SS-1), therefore, the reaction mixture showed a self-setting property as free Ca^{2+}, $H_2PO_4^-$, and HPO_4^{2-} ions were gradually consumed, and the growth of monetite crystals took place according to the tentative reaction:

$$Ca(OH)_2 + H_3PO_4 + Na^+(aq) + HCO_3^-(aq) \rightarrow CaHPO_4 + 2H_2O + Na^+(aq) + HCO_3^-(aq) \quad (1).$$

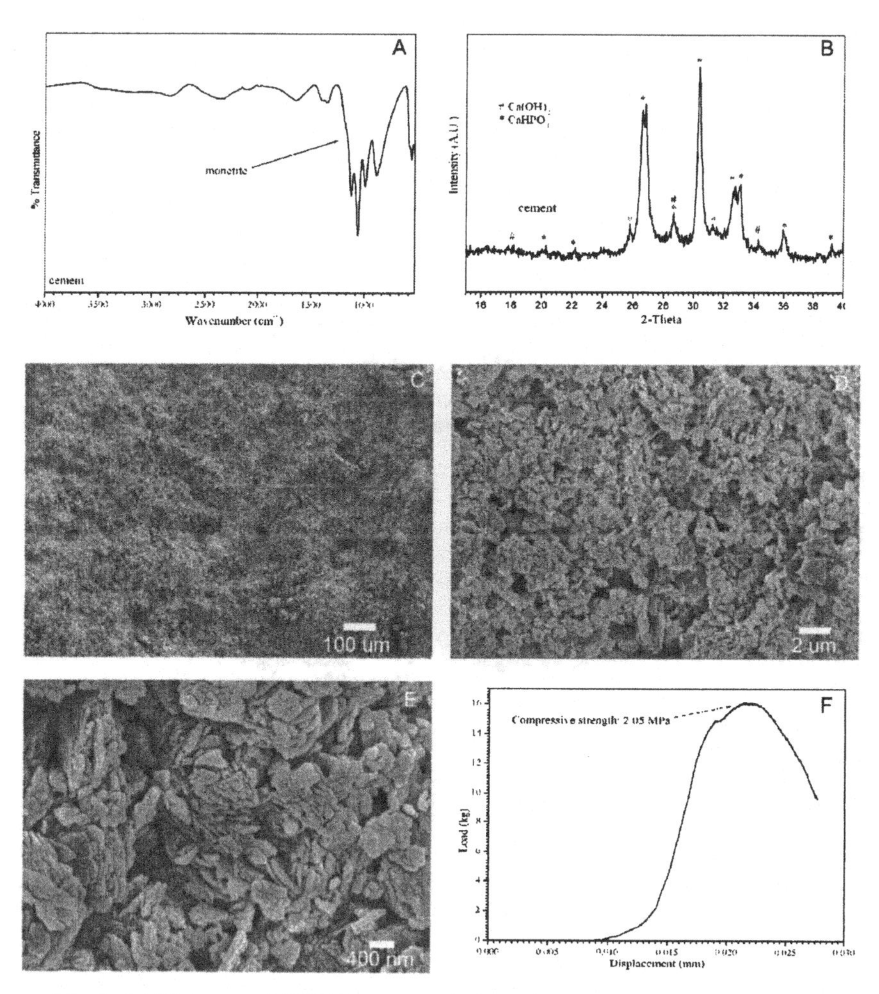

Fig. 3 (*a*) IR spectra of $CaHPO_4$ cement, (*b*) XRD data of $CaHPO_4$ cement, (*c*), (*d*) and (*e*) SEM images of $CaHPO_4$ cement, (*f*) compressive strength data of $CaHPO_4$ cement

Although the above reaction showed the use of H_3PO_4 in the reactants, the setting solution (SS-2) which gave rise to the synthesis of this inexpensive and low-strength $CaHPO_4$ cement presumably contained a mixture of $H_2PO_4^-$ and HPO_4^{2-} ions, together with some hydrogen carbonate ions. MCPM is known to be the stable phase at a pH below 2.5, whereas monetite ($CaHPO_4$) becomes the stable phase over the pH range of 2.5-4.2[23].

The addition of a small amount of $NaHCO_3$ to the H_3PO_4-based setting solution was quite influential in bringing the pH of the cement paste down to a range where monetite became the most favored phase, and this has been the main idea behind this very simple cement. In vitro testing of these cements is in progress.

CONCLUSIONS

A very simple, inexpensive and a robust technique for the preparation of a cement consisting of single phase monetite ($CaHPO_4$) was presented. $Ca(OH)_2$ was reacted with an aqueous mixture of H_3PO_4 and $NaHCO_3$ to form $CaHPO_4$ in solid, cement form. Since the final product of the cement was only $CaHPO_4$ of high aqueous solubility, the cement bodies are also regarded to have high bioresorbability in comparison to apatitic cements. The cement pastes showed very good handling properties and possessed a setting time less than 20 minutes.

ACKNOWLEDGMENTS

This study was funded in part by NSF 0522057.

REFERENCES

[1]L. C. Chow and W. E. Brown, "Dental Restorative Cement Pastes," US Patent No. 4,518,430, (1985).

[2]L. C. Chow and W. E. Brown, "A New Calcium Phosphate Setting Cement," *J Dent Res*, **63**, 672-6 (1983).

[3]B. R. Constantz, B. M. Barr, I. C. Ison, M. T. Fulmer, D. C. Delaney, J. Ross, and R. D. Poser, "Histological, Chemical and Crystallographic Analysis of Four Calcium Phosphate Cements in Different Rabbit Osseous Sites," *J. Biomed. Mater. Res.*, **43**, 451-61 (1998).

[4]E. M. Ooms, J. G. C. Wolke, J. P. C. M. van der Waerden, and J. A. Jansen, "Use of Injectable Calcium Phosphate Cement for the Fixation of Titanium Implants: An Experimental Study in Goats," *J. Biomed. Mater. Res.*, **66B**, 447-56 (2003).

[5]S. Jalota, A. C. Tas, and S.B. Bhaduri, "Synthesis of HA-seeded TTCP ($Ca_4(PO_4)_2O$) Powders at 1230°C from $Ca(CH_3COO)_2{\cdot}H_2O$ and $NH_4H_2PO_4$," *J. Am. Ceram. Soc.*, **88**, 3353-60 (2005).

[6]L. C. Chow, M. Markovic, S. A. Frukhtbeyn, and S. Takagi, "Hydrolysis of Tetracalcium Phosphate under a Near-Constant-Composition Condition—Effects of pH and Particle Size," Biomaterials, 26, 393–401 (2005).

[7]Y. Matsuya, S. Matsuya, J. M. Antonucci, S. Takagi, L. C. Chow and A. Akamine, "Effect of Powder Grinding on Hydroxyapatite Formation in a Polymeric Calcium Phosphate Cement Prepared from Tetracalcium Phosphate and Poly(methyl vinyl ether maleic acid)," *Biomaterials*, **20**, 691-7 (1999).

[8]S. R. Radin and P. Ducheyne, "Effect of Bioactive Ceramic Composition and Structure on in-Vitro Behavior. 3. Porous versus Dense Ceramics," *J. Biomed. Mater. Res.*, **28**, 1303-9 (1994).
[9]A. C. Tas, "A New Calcium Phosphate Cement Composition and a Method for the Preparation Thereof," US Patent No. 6,929,692 (2005).
[10]A. A. Mirtchi, J. Lemaitre and N. Terao, "Calcium-Phosphate Cements - Study of the Beta-Tricalcium Phosphate - Monocalcium Phosphate System," *Biomaterials*, **10**, 475-80 (1989).
[11]K. Ohura, M. Bohner, P. Hardouin, J. Lemaitre, G. Pasquier, and B. Flautre, "Resorption of, and Bone Formation from, New Beta-tricalcium Phosphate-Monocalcium Phosphate Cements: An In Vivo Study," *J. Biomed. Mater. Res.*, **30**, 193-200 (1996).
[12]M. Bohner, H. P. Merkle, P. V. Landuyt, G. Trophardy, and J. Lemaitre, "Effect of Several Additives and Their Admixtures on the Physico-chemical Properties of a Calcium Phosphate Cement," *J. Mater. Sci. Mater. M.*, **11**, 111-6 (1999).
[13]M. Bohner, F. Theiss, D. Apelt, W. Hirsiger, R. Houriet, G. Rizzoli, E. Gnos, C. Frei, J. A. Auer, and B. von Rechenberg, "Compositional Changes of a Dicalcium Phosphate Dihydrate Cement after Implantation in Sheep," *Biomaterials*, **24**, 3463-74 (2003).
[14]G. Penel, N. Leroy, P. Van Landuyt, B. Flautre, P. Hardouin, J. Lemaitre, and G. Leroy, "Raman Microspectrometry Studies of Brushite Cement: In Vivo Evaluation in a Sheep Model," *Bone*, **25**, 81S-84S (1999).
[15]M. Bohner, U. Gbureck, and J. E. Barralet, "Technological Issues for the Development of More Efficient Calcium Phosphate Bone Cements: A Critical Assessment," *Biomaterials*, **26**, 6423-9 (2005).
[16]U. Gbureck, S. Dembski, R. Thull, and J. E. Barralet, "Factors Influencing Calcium Phosphate Cement Shelf-life," *Biomaterials*, **26**, 3691-7 (2005).
[17]S. Takagi and L. C. Chow, "Self-Setting Calcium Phosphate Cements and Methods for Preparing and using Them," US Patent No. 5,525,148 (1996).
[18]E. Fernandez, F. J. Gil, M. P. Ginebra, F. C. M. Driessens, J. A. Planell, and S. M. Best, "Calcium Phosphate Bone Cements for Clinical Applications - Part I: Solution Chemistry," *J. Mater. Sci. Mater. M.*, **10**, 169-6 (1999).
[19]L. C. Chow, S. Takagi, R. J. Shern, T. H. Chow, K. K. Takagi, and B. A. Sieck, "Effects on Whole Saliva of Chewing Gums containing Calcium Phosphates," *J. Dent. Res.*, **73**, 26-32 (1994).
[20]D. Bayraktar and A. C. Tas, "Chemical Preparation of Carbonated Calcium Hydroxyapatite Powders at 37°C in Urea-containing Synthetic Body Fluids," *J. Eur. Ceram. Soc.*, **19**, 2573-9 (1999)
[21]R. Z. LeGeros, J. P. LeGeros, O. R. Trautz, and W. P. Shirra, "Conversion of Monetite, $CaHPO_4$, to Apatites: Effect of Carbonate on the Crystallinity and the Morphology of the Apatite Crystallites," *Adv. X-ray Analy.*, **14**, 57-66 (1971).
[22]F. Barrere, C. A. van Blitterswijk, K. de Groot, and P. Layrolle, "Influence of Ionic Strength and Carbonate on the Ca-P Formation from SBFx5 Solution," *Biomaterials*, **23**, 1921-30 (2002).
[23]E. Fernandez, F. J. Gil, M. P. Ginebra, F. C. M. Driessens, J. A. Planell and S. M. Best, "Calcium phosphate bone cements for clinical applications - Part II: Precipitate formation during setting reactions," *J. Mater. Sci. Mater. M.*, **10**, 177-83 (1999)

CHEMICALLY BONDED CERAMICS BASED ON CA-ALUMINATES AS BIOMATERIALS

L. Hermansson[1,2] and H. Engqvist[1,2]
[1]Doxa AB
[2]Materials Science Department, The Angstrom Laboratory, Uppsala University
leif.hermansson@doxa.se; hakan.engqvist@angstrom.uu.se

ABSTRACT

This presentation aims at giving an overview picture of the potential of Ca-aluminate as biomaterial. Identified possible applications for Ca-aluminate based materials are within vertebroplasty and odontology. An additional field of interest is as an implant coating material. Ca-aluminate with oxide particles as well as Ca-aluminate with glass particles were examined concerning handling and mechanical properties, biocompatibility and bioactivity. For both orthopaedic and dental injectable CA-pastes the working time of 4 minutes and the setting time about 10 minutes are achievable. The fracture toughness was determined to be in the range 0.5-1.0 $MPam^{1/2}$ and the compressive strength in the interval 150-250 MPa depending on selected w/c ratio, and the flexural strength above 50 MPa. The hardness is about 100 HV, and Young's modulus 10-15 GPa. The hydrates formed are in the size range of 20-50 nm. In hard tissue the Ca-aluminate based materials form a chemical interface totally closing the gap between the biomaterial and the surrounding tissue. Two features contribute to this sealing of interfaces; first the general dissolution of Ca-aluminate and precipitation of nanosize hydrates in micro-voids between particles and tissue, and on all possible walls including tissue surfaces, and second the zone formation including possible apatite formation upon which bioactive-induced formation of new tissue including apatite formation. Thus both a pure chemical integration and a biologically induced integration contribute to sealing of biomaterial-tissue interfaces.

INTRODUCTION

Inorganic cements are gaining increased attention as biomaterials. These chemically bonded biomaterials comprise inorganic cements from chemical groups such as phosphates, silicates, aluminates, sulphates and carbonates. Biocements are often based on various calcium phosphate salts – often in combination with Ca-sulphates. These salts can be made to cure *in vivo* and are attractive as replacements for the natural calcium phosphates of mineralised tissues. The calcium phosphate products are gaining ground in orthopaedics as resorbable bone substitutes. However, these products have low mechanical integrity (compression strength values in the interval 10-40 MPa) and may be questioned as load bearing implants [1-3].

Materials based on Ca-aluminates and also Ca-silicate with chemistry similar to that of Ca-phosphates contribute to some additional features of interest with regard to dental and orthopaedic applications. These features are related to the amount of water involved in the curing process, the early and high mechanical strength obtained, and the biocompatibility profile including *in situ* reactions with phosphates ions of the body fluid [4-6].

The examination in this paper deals with aspects of chemical reactions at different stages, initial hydration, long-term stability and reactions in tissue environment. Properties dealt with in this survey comprise handling, mechanical properties, dimensional stability, biocompatibility and bioactivity. In Table I general aspects behind the selection of Ca-aluminate based materials as biomaterials are presented.

Table I. The rationale behind the use of Ca-aluminates as biomaterials [4].

Chemistry	Composed of frequently occurring elements, oxides, hydrophilic nature, similarities with apatite
Biology	Biocompatible (also during curing) and environmentally friendly
Thermal	Expansion and conductivity comparable to hard tissue
Mechanics	Hardness and stiffness comparable to hard tissue, bone, strong in compression
Processing	*In situ* room-temperature preparation, adjustable rheology and curing time

EXPERIMENTAL

Materials.

The raw materials used were of different origin, namely one commercially available Ca-aluminate powder, Secar 71 (LaFarge), and Ca-aluminate materials synthesised by Doxa AB. Secar 71 contains the Ca-aluminate phases $CaOxAl_2O_3$ (CA) and $CaOx2Al_2O_3$ (CA_2), while the Doxa Ca-aluminate is based on a pure phase CA or 12 $CaOx7Al_2O_3$ ($C_{12}A_7$). The raw materials were milled to obtain a fine-grained particle size distribution. The different solids added were oxides or silicate glass depending on the aimed application area, orthopaedics or dentistry. In cases where the chemically bonded ceramics are used for surface modification of implants no additional phase was included besides Ca-aluminate. The paste for coating purpose is composed of Grossite ($CaO{\cdot}2Al_2O_3$) and an isostructural form of Marokite, where Fe_2O_3 is exchanged for Al_2O_3, i.e. $CaO{\cdot}Al_2O_3$. The ratio of calcium- to aluminium oxide is approximately 2/3. The inert phase in the orthopaedic injectable material was zirconia.

Preparation of the final powder compositions includes mixing by blending the constituents using silicon nitride balls in a polyethylene container with iso-propanol. The alcohol was evaporated and polymeric residues were removed in a furnace at 400 °C. The reaction liquid is water. The powder mix was sterilized by electron beam radiation, and the aqueous solutions were steam sterilized.

The storage solutions employed were pure water and also a phosphate buffer solution (PBS). All tests were conducted at 37 degree C. Test samples for orthopaedic applications also included different processing agents for dispersive and viscosity control. In all bulk test samples an accelerator of a Li-salt in concentration of 5-30 ppm was used.

Reference materials in the studies were commercial PMMA-based materials and Ca-phosphate based materials.

Methods and property valuation.

Properties dealt with were working, setting and injection time, radiopacity, mechanical properties such as facture toughness, compressive and flexural strength, hardness and Young's modulus, dimensional stability aspects during hardening, biocompatibility (cell culture studies and animal models) and bioactivity (formation of apatite layer *in vitro* and bonding to tissue *in vivo*). The *in vitro* bioactivity tests were conducted according to the outline in the ISO-standard ISO/WD 23317 This standard describes a procedure for producing simulated body fluid (SBF), sample preparation, immersion and analysis techniques. For analyses of possible apatite formation scanning electron microscope (SEM), energy dispersive X-ray spectroscopy (EDX) and X-ray diffraction (XRD) were used.

To analyse interfaces and calcified tissue at the highest level transmission electron microscopy (TEM) was used in combination with focused ion beam microscopy (FIB) for intact site-specific preparation of the TEM-samples. To produce the TEM samples in this investigation the so-called "H-bar lift-out" technique was used enabling a very high site-specific accuracy [7,8]. The thickness of the TEM samples in this investigation was about 150 nm. The site-specific accuracy of the technique is about 1 μm. FIB was used also for high-resolution imaging. The interface elemental and phase analyses were determined using energy dispersive X-ray mapping and electron diffraction and imaged in scanning-TEM and bright field mode. The technique fulfils a demand to correlate the surface properties of implants with the structure and composition of preserved interfaces with tissues.

In a cohesiveness test [9] of the material one ml of Ca-aluminate cement paste was extruded into the different media at 3 minutes and 6 minutes after mixing. Images of the extruded material in the media were taken after the material was set, i.e. > 15 minutes, using a digital camera. To analyse the composition of the "cloud" surrounding the material, the "cloud" was collected from the media using a syringe. The samples were evaporated at 110°C in glass beakers. The dried powders were collected on carbon tape and analysed with SEM/EDX.

Blood contact evaluation was conducted according to a system described in more details in [10]. The materials were tested with a closed circuit Chandler loop model. In this test, the blood circulation is simulated with a rotating loop of PVC tubing with a total length 500 mm and an inner diameter of 4 mm, making a total inner volume of 6.3 ml. The inner surface of the PVC tubing is coated with an immobilized functional heparin. After application of the test materials and 4.5 ml (leaving a small air pocket facilitating the flow) of fresh blood, the closed tubing loops are rotated for one hour at 32 rpm in a 37°C water bath. Thereafter the blood and test materials are collected and investigated. For comparison reason commercial products based on PMMA and Ca-phosphate and Ca-sulphate were included in the clotting test.

Threaded screw implants of pure titanium (grade 2), dental implants with a modified head, were used in the coating experiments [11]. The length of the threaded (non-conical) part is 4 mm and its outer diameter 3.75 mm. The implants were evaluated as machined or with either of two types of calcium aluminate coatings. Implants were also anchored with a calcium aluminate paste. Coatings of about 30-50 μm thickness were produced on the threaded part of the Ti-implants with flame spraying equipment using a calcium aluminate powder of the $(CaO)_{12}(Al_2O_3)_7$ phase. Prior to spraying the samples were blasted with alumina grit to a surface roughness of about 2 μm. During deposition the substrates reached about 70 °C. Coatings were also deposited with sputtering using a high-vacuum system with a background pressure of 10^{-8} Torr using a radio-frequency power source and powder feeds on the magnetron of about 100 W. A planar $(CaO)_{12}(Al_2O_3)_7$ calcium aluminate multiple-piece target was used. The target sections were sintered from crushed and sieved powder of about 10 μm grain size. The films were grown on the Ti-implants, directly facing the substrates at a target-to-substrate distance of 7 cm.

The animal model to test the materials involved female albino adult New Zealand White rabbits. The paste was applied to the Ti-implants by dipping the implants in the paste. The implants were screwed into pre-drilled holes (diameter 4 mm and length 8 mm). The paste was also injected into pre-drilled holes, 5 mm in diameter and approximately 12 mm deep, in the tibia condyle. Autopsy took place after 6 weeks. Sections including both the bone and the implant were cut. The surgical procedure followed standard techniques. Polymethylmetacrylate (PMMA) based bone cement (CMW 1 from Johnson & Johnson) was used as reference material. After necropsy and stabilisation in formaldehyde, the samples were imbedded in an acrylate polymer, cut and polished. The sample preparation procedure principally involves: Dehydration for one week in 70 %, 95 % and 98 % ethanol, respectively; infiltration for one week in glycolmethacrylate/alcohol mixes of increasing concentrations, ending with pure glycolmethacrylate, respectively; and finally polymerisation with UV-light. The specimens were split with a diamond saw blade and the cut surface polished [11].

RESULTS

The results will be presented in three major parts related to hydration (initial hydration and curing), long-term stability and interaction with living tissue.

Initial hydration and curing.

Below are presented the main chemical reactions controlling the curing of Ca-aluminate as well as the property profile developed during the hydration process. Also the (dis)/integration property (the cohesiveness) of the material in contact with water and body solution is presented as well as the pH development and clotting behaviour (blood contact interaction), all important from biocompatibility and bioactivity perspectives, are discussed.

The Ca-aluminates react with water-containing solution according to the following stages and reactions at temperatures > 30 ^{0}C [4]; Ca-aluminates dissolution in water forming the ions Ca^{2+}, $Al[OH]_4^-$ and OH^-, saturation of ions in the liquid and precipitation of the phases kaotite and gibbsite.

$$3\,(CaO\ Al_2O_3) + 12\ H_2O \rightarrow 3\ Ca^{2+} + 6\ Al^{3+} + 24\ OH^- \rightarrow$$
$$3\ Ca^{2+} + 6\ Al(OH)_4^- \rightarrow$$
$$Ca_3\ [Al(OH)_4]_2(OH)_4\ \text{(katoite)} + 4\ Al(OH)_3\ \text{(gibbsite)} \qquad (1)$$

This precipitation is repeated until the reaction is completed, i.e all Ca-aluminate is consumed, or incomplete due to diffusion-hindrance and/or lack of water, the latter depending on the w/c ratio selected. The on-going hydration in the initial stage determine, the working and setting time, as well as the time-dependant strength development. The hydration is fast during the first hours, and most of the hydration occurs during the first day. After a week possible further hydration is difficult to detect. The setting time of the material could be controlled to be in the interval 3-10 minutes by changing dispersant agents and accelerator concentration. For dispersant reasons a methyl metacellulose was used. Typical properties are presented in Tables II-III. In all these tests at least 8 samples were tested for each condition.

Table II. Hardness, compressive strength and residual porosity as a function of time of an experimental injectable Ca-aluminate based material.

Property	**After 1h**	**After 24 hrs**	**After 7 days**	**After 28 days**
Hardness (MPa)	30	60	95	100
Compressive strength (MPa)	20	70	105	105
Porosity (%)	21	10	6	6

A typical property profile of a high-strength Ca-aluminate based material with low w/c ratio is shown in Table III.

Table III. Some property data (mean values) of a Ca-aluminate based material with a w/c ratio of 0.32 [12,16].

Property	Mean value
Hardness (Hv 100 g) after 28 days	110
Compressive strength (MPa) after 28 days	258
Flexural strength (MPa) after 7 days	81
Fracture toughness, $MPam^{1/2}$	0.8
Initial setting (min)	3
Young´s modulus, GPa	15
Final setting (min)	6

During the hydration process pH is alkaline in the interval 8-11.5 with decreasing pH towards neutral as the curing proceeds [13]. The alkaline situation makes also possible precipitation of biominerals such as apatite and carbonate possible, especially in contacts areas towards tissue. High pH contributes to the following changes in body liquid

$$H_2PO_4^- + OH^- \rightarrow HPO_4^- + H_2O \qquad (2)$$

$$HPO_4^- + OH^- \rightarrow PO_4^{3-} + H_3O \qquad (3)$$

and present ions contribute to possible direct apatite formation early in the hydration process in contact with body liquid

$$5\,Ca^{2+} + 3\,PO_4^{3-} + OH^- \rightarrow Ca_5 \cdot (PO_4)_3 \cdot (OH)\,. \qquad (4)$$

The pH development during curing is presented in Table IV. Both the development in pure water and in a saliva solution is shown.

Table IV. The pH development during hydration of Secar 71 (1h measurements within brackets) [14].

Sample	1 WC	2 W	3 W	4 SC	5 S
1h	11.54	11.38	11.29	10.30	10.24
24h	11.78	11.55 (9.37)	11.42 (9.78)	10.68	10.23 (7.41)
7 days	11.22	11.26 (8.77)	11.42 (8.58)	10.34	9.92 (7.82)
14 days	10.17	10.12 (8.97)	10.28 (8.98)	9.89	9.54 (8.08)
28 days	10.20	10.51 (8.42)	10.45 (8.25)	9.80	9.25 (8.13)

WC = in distilled water continuously, W = in distilled water exchanged at every measurement occasion, SC = in saliva continuously, S = in saliva exchanged at every measurement. occasion.

In the soft (adipose) tissue contact zone some areas of calcite, $CaCO_3$, have been found, formed in the initial basic environment according to:

$$HCO_3^- + Ca^{2+} + OH^- \rightarrow CaCO_3 + H_2O. \qquad (5)$$

The cohesiveness evaluation revealed differences with regard to the solution used. In tap water a cloud was formed around the extruded material (3 minutes after mixing), see Fig. 1a. A smaller cloud formed when the cement was extruded into tap water at 6 minutes after mixing. In PBS, the cloud seems to consist of agglomerates or flakes, see Fig. 1b.

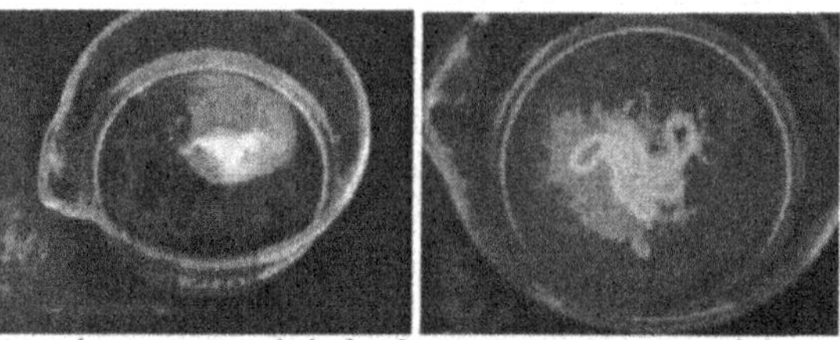

Fig. 1: Photos of calcium aluminate extruded after 3 minutes into tap water (left) and PBS (right). [9]

Regarding the composition of the cloud, the cloud formed in water showed traces of calcium aluminate constituents, see Fig. 2a. Note especially the presence of Zr (added in the form of ZrO_2 for radiopacity reason). ZrO_2 is insoluble in water and thus only appeasr as grains. For the test in PBS, ions from the PBS was present in the precipitates, see Fig. 2b. Also, on the surface of the set cement P could be found, indicating apatite formation.

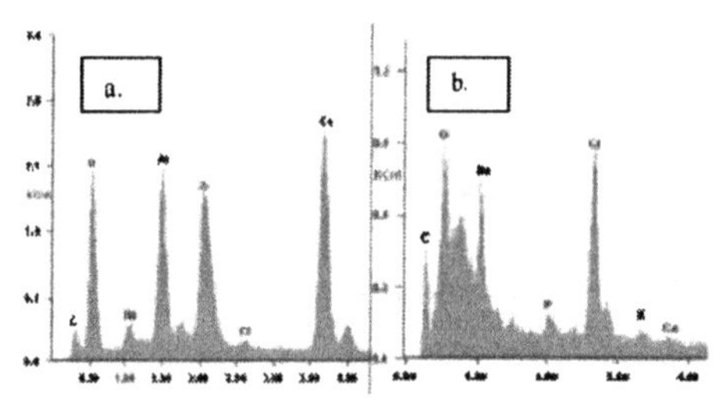

Fig. 2: Elemental composition of the collected powder (SEM/EDS) a) water and b) PBS.

The blood response evaluated as the clotting behaviour and platelet count of the Ca-aluminate cement and the PMMA was comparable in that both materials showed very low clotting tendency and were able to maintain high platelet count numbers, around 200 for both materials. The calcium phosphate and sulphate materials were strikingly different and both showed strong clotting as well as low platelet counts, around 10, see Table V.

After centrifugation, no or low signs of haemolysis were found for the Ca-aluminate cement and the PMMA, whereas the calcium phosphate and calcium sulphate materials produced some haemolysis, as revealed by a reddish colour of the plasma. Hence both the visual indications and the platelet count result show that calcium phosphate and calcium sulphate materials produced strong clotting in these tests. The Ca-aluminate material (CA) and PMMA both produce low clotting and the platelet count values are comparable to the control.

Table V. Blood analysis after termination of the Chandler loop test [10].

Material	Platelet count	C3a (μg/ml)	TCC (AU/ml)	TAT (pmol/ml)
Ca-aluminate based	198-271	628-751	137-481	7.4-29.0
Ca-phosphate based	8-37	-	-	-
Calcium sulphate	4-14	-	-	-
PMMA	217-308	303-444	109-126	3.4-11.6
Control	144-304	152-317	30-39	2.6-40.9
Baseline	229-335	34-417	8.5-12	2.1-3.6

Also, the control loops did not induce any complement activation as reflected by formation of C3a. Both PMMA and Ca-aluminate induced moderate activation of the C3a component of the complement. The Ca-aluminate biocement displayed somewhat higher tendency in this respect. Also the control had a slightly increased TCC value compared to the baseline. The formation of the soluble terminal complex of complement, TCC was slightly elevated in the control loops. This has previously been shown to be linked to the presence of an air interface. The TCC data obtained from the loops with PMMA and Ca-aluminate biocement were consistent with the data obtained on C3a.

Long-term stability aspects

Typical microstructures of cured Ca-aluminate based materials are shown in Figs. 3-5. An additional aspect of the microstructure obtained is that for larger grains a conversion within the grain may occur, Fig. 4.

Figure 3. SEM micrograph of an experimental Ca-aluminate material for odontological applications.. The small white spots are glass particles, the light-grey phase katoite ($Ca_3[Al(OH)_4]_2(OH)_4$ and the dark-grey phase gibbsite ($Al(OH)_3$).

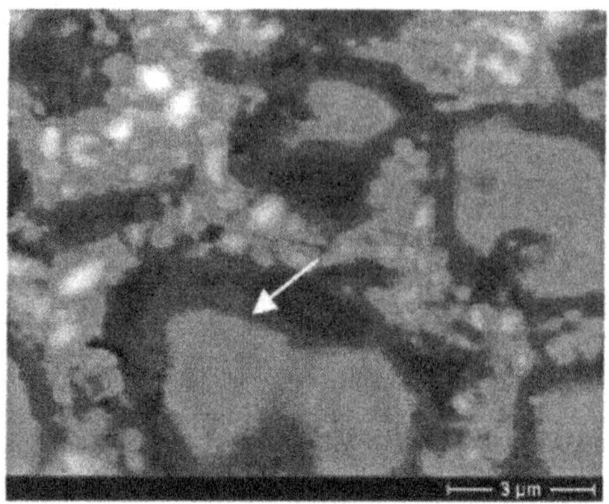

Fig.4 Hydration within larger Ca-aluminate grains (white arrow)

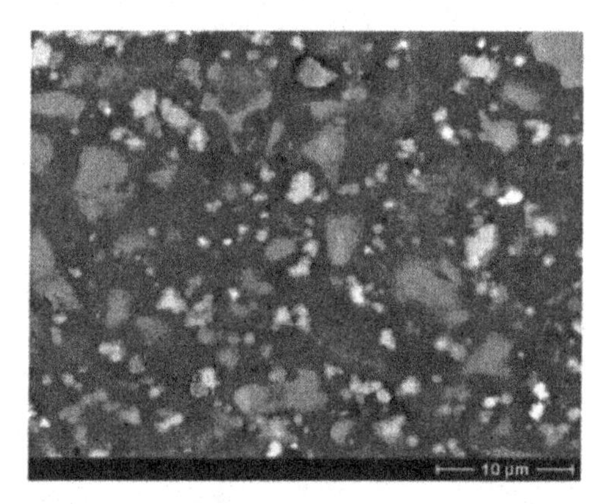

Figure 5. SEM micrograph of an experimental Ca-aluminate paste for orthopaedic applications microstructure. a) general microstructure, The white spots are oxide particles.

High resolution transmission microscopy (HRTEM) of hydrated Ca-aluminate paste reveals the hydrated phases to be very well connected – on the crystal lattice distance. The size of the hydrates is estimated to be in the interval 20-50 nm. See Fig.6. This nano-size range of hydrates was proposed by Power and Brownyard as early as 1946, based on BET measurements of cement [15].

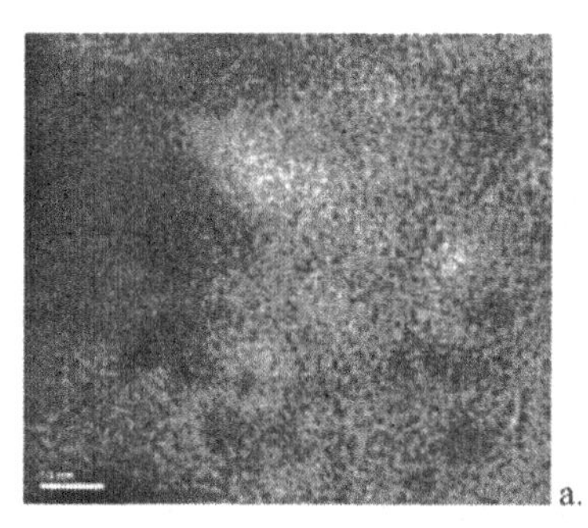
a.

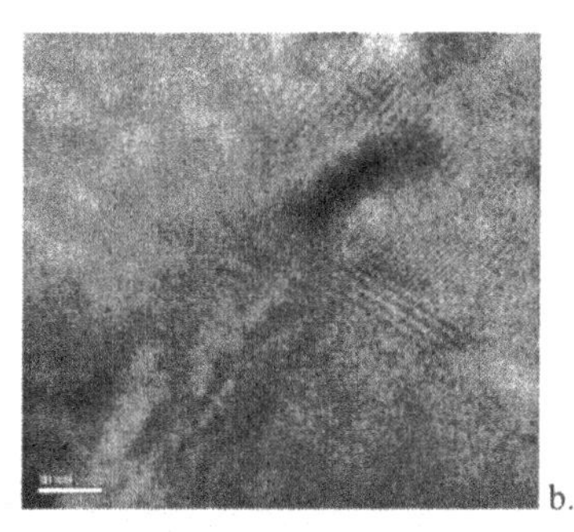
b.

Fig 6. High resolution TEM images of the implanted Ca-aluminate cement paste showing the lattice (white bar = 10 nm)

Studies of the dimensional stability and continued water-up take due to the hydration reaction show that after some weeks the reaction can be judged as complete. The dimensional change during hydration and curing can be controlled to be below 0.3 linear %. The corresponding expansion pressure can be kept below 3 MPa [16]. To obtain these low levels of dimensional change the use of expansion-controlling additives are essential [4].

The Ca-aluminate materials are chemically resistant within the pH interval 2.5 – 11. At pH 2.7 the Ca-aluminate system does not show any dissolution at all in the ISO-9917 acid erosion test. However, a zone formation when placed in different media has been detected. This is summarised below, and the controlling step is the solubility of possible phases formed. While many polymer systems often repel water, the chemically bonded ceramics are strongly hydrophilic and are easily mixed with the water of surrounding body fluids and tissues. This contributes to the formation a surface zone. Water diffusion increases the water to powder ratio to a certain depth, and the precipitation of phases during hydration is affected by the ions being present in the tissue liquids, primarily calcium-, phosphate-, and carbonate ions. Also the increased pH created during the hydration favours the precipitation of phases stable in alkaline solutions, i.e. apatite, gibbsite and calcite. The bulk reaction occurs according to equation 1 above. Formed Al-ions in the initial hydration dissolution stage are instantly transferred to aluminate ions, $Al(OH)_4^-$ in the basic environment, and consumed (precipitated) as katoite and gibbsite. During this curing time in the contact zone to body tissue, apatite is formed in addition as the more stable phase. Katoite has a small but higher solubility than that of apatite. Thus it can be hypothesized [13] that katoite is dissolving at the surface in a Ca^{2+} and phosphate ion acid-base mediated reaction according to:

$$Ca_3 \cdot (Al(OH)_4)_2 \cdot (OH)_4 + 2\,Ca^{2+} + HPO_4^{2-} + 2\,H_2PO_4^- \rightarrow$$

$$Ca_5 \cdot (PO_4)_3 \cdot (OH) + 2\,Al(OH)_3 + 5\,H_2O \qquad (6)$$

In this reaction apatite and gibbsite are produced. The katoite, gibbsite and apatite phases are found in agreement with the chemical reactions (1, 6) in the surface contact zone. See Figs. 7-

8 below. This zone seems to keep its thickness over time. The tendency for the contact zone formation is high during the initial hydration phase, where unreacted Ca-aluminate is present. However, the proportion of this unreacted phase is reduced as the initial hydration continues. After approximately one month no activity is taking place with regard to the Ca-aluminate curing (hydration) [13]. During this time period calcium and phosphate ions from the extracellular fluid help to form apatite in the contact zone. Thereafter the activity in the contact zone is minimal. The physiological buffer system keeps the pH at near neutrality. During these circumstances the general dissolution of the system is extremely small.

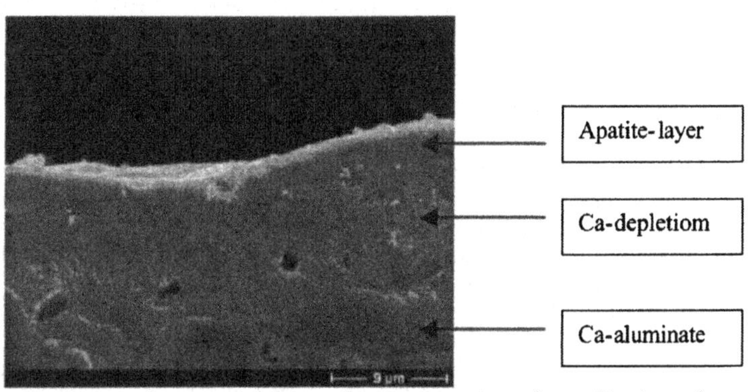

Fig. 7. SEM micrograph of apatite layer on a calcium aluminate specimen formed in vitro after storage in phosphate buffer solution for 4 months. Ca-depletion beneath the apatite-layer.

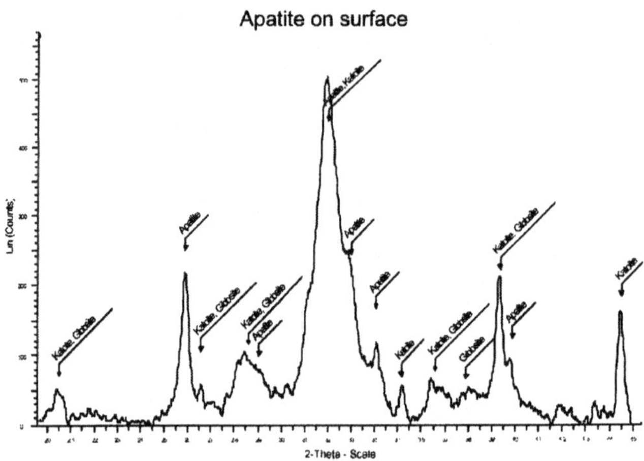

Fig. 8. X-ray diffraction pattern from apatite layer formed on a calcium aluminate specimen in vitro after storage in phosphate buffer solution for 4 weeks.

<u>The contact zone between Ca-aluminate materials and hard tissue.</u>

The interface between the surface zone and new hard tissue formed has been studied in some details using SEM and high-resolution TEM-FIB. In these analyses total closure of the gap between the biomaterial and tissue has been found. As examples the situation for a dental application and an orthopaedic application are presented in Figs. 9-10.

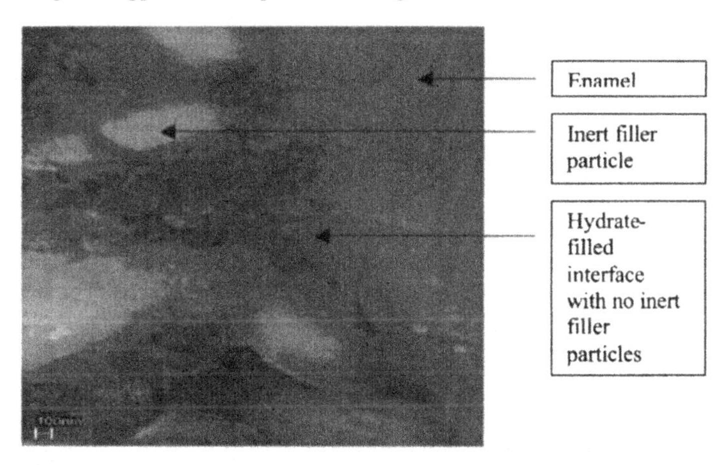

Fig 9. SEM micrograph of the contact zone between a Ca-aluminate restorative material and enamel, white bar = 100 nm [17].

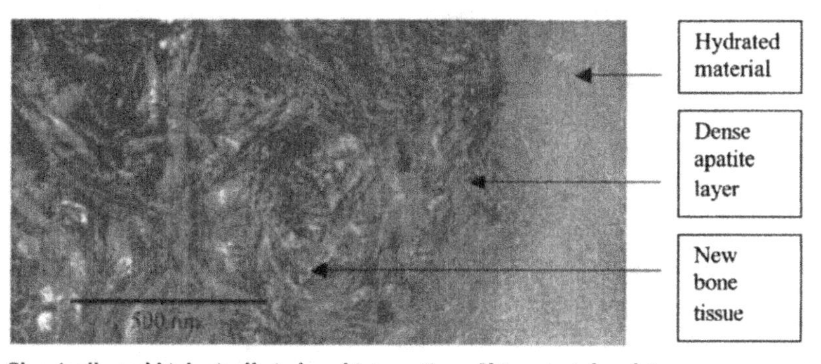

Fig. 10 Chemically and biologically induced integration of biomaterial and tissue

Two features contribute to this perfect sealing of interfaces; first the general dissolution of Ca-aluminate and precipitation of nanosize hydrates in micro-voids between particles and between particles and tissue, and on all possible walls including tissue surfaces, and second the zone formation including possible apatite formation upon which bioactive- induced formation of new tissue including apatite formation. Thus both a pure chemical integration and a biologically induced integration contribute to sealing of biomaterial-tissue interfaces. This inherent property may also have a major impact on the marginal integrity of a restoration leading to a better seal of the cavity and thus reduction of the incidence of micro-leakage as discussed in a recent paper [17].

Optimal early fixation of implants to bone tissue is instrumental both in allowing for early loading of the implant, and to reduce the risk of long-term loosening resulting from micro-motions. The use of surface coatings technology is today an established method to reduce the problem with poor interfacial stability of implants. With coatings technology, structural characteristics of the implant bulk material (e.g. strength, ductility, low weight or machinability) may be combined with surface properties promoting tissue integration. There are several established coating deposition techniques, e.g. physical vapour deposition (sputtering) and thermal spraying techniques. Coatings based on calcium phosphates are much used, most commonly they are deposited with thermal spraying techniques.

The present study also deals with prospects of achieving early anchoring of metal implants by exploring coatings based on chemically bonded ceramics, particularly coatings of calcium aluminate, which were allowed to hydrate *in vivo.* Results are presented from an implantation study with coated titanium implants and uncoated implants augmented with a calcium aluminate paste in the hind legs of rabbits.

No negative effects of the implants on the general welfare of the animals were observed. The healing progressed in a normal and favourable way. As for the removal toque recordings, all calcium aluminate coatings types provided an improved implant anchoring to bone tissue after *in vivo* hydration as compared to that of the pure metal implants. Implants on the tibia and femur side of the knee gave similar removal torques. Table VI provides average values from both the tibia and femur sides. The sputtered implant gave the highest removal torque, although for these implants the statistics is insufficient (n=3) for any certainty in a conclusion. After 2 weeks of implantation, implants combined with paste augmentation provide the highest removal torque, and also flame sprayed coatings improve the torque relative to the uncoated system. Sputtered coatings lack for this time interval. After six weeks of implantation, all systems are relatively similar (considering the uncertainty due to scatter and statistics), apart from the sprayed system, which shows significantly higher values, see Table VI.

Table VI. Removal torque (Ncm) for dental implants in rabbit hind legs (tibia and femur) [11].

Implant type	**24 hrs**	**(n)**	**2 weeks**	**(n)**	**6 weeks**	**(n)**
Flame spraying	7.0	(8)	7.0	(8)	25	(6)
Paste augmentation	6.6	(8)	15	(6)	13	(4)
Sputtering	12	(3)	-	-	10	(2)
Uncoated reference	3.8	(8)	5.7	(6)	14	(4)

The bone regeneration process developed favourably around the implants and after six weeks a natural healing involving a densification of the trabecular bone around the implants was seen. No signs of unfavourable healing due to the presence of hydrated calcium aluminate were found. The paste was found to be unevenly distributed around the implants, see pink areas in Fig. 11.

Fig. 11 Dental implant with sputtered calcium aluminate after 6 weeks of implantation in the femur of a rabbit (left), and dental implant anchored with calcium aluminate paste after 24 hrs of implantation (right) [11].

The results indicate that calcium aluminate coatings provide adequate adherence to the substrate. In the torque tests, the implant to bone interface proved the weakest interface. In general, there was a good apposition between calcium aluminate and cortical bone. There were no obvious inflammatory reactions.

Contact zone between Ca-aluminate and substrate

HRTEM of the contact zone between the hydrated paste and the implant substrate surface shows the interface to be gap-free. This contact is so close that it can be claimed to be on the atomic level. See Fig. 12.

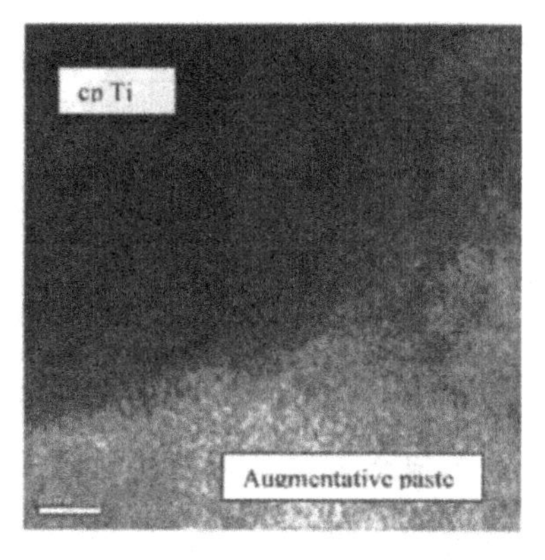

Fig. 12 HRTEM of the contact zone between the augmentative paste coating and cp Ti (white bar = 10 nm).

Also the reported test results from the torque evaluation above show that the fractures very seldom occur in the interface to the substrate but rather in the surrounding hard tissue or in the interface between hard tissue and the biomaterial.

DISCUSSION

There are several chemically bonding ceramic systems, mainly calcium based; e.g. Ca-silicates, Ca-aluminates, Ca-sulphates and Ca-phosphates. The chemistry of these materials is similar to that of hard tissue in living organisms, the latter being based on apatites and carbonates. Chemically bonded ceramics can be bioactive and generally display great potential as biomaterials. Calcium aluminates have advantages in terms of mechanical strength resulting from a high turn-over of water during hydration and related pore closure.

The Ca-aluminate based materials show from a materials science viewpoint high potential in biomaterials applications, where now organic polymer-based materials and Ca-phosphate based materials are used within dentistry and orthopaedics, due to some complementary properties, especially related to early high strength and promising biocompatibility. In Table VII the property profile of CA-based materials is compared to those of commercial materials.

Table VII . Typical properties of Ca-aluminate, PMMA and Ca-phosphates.

Property (typical)	**Ca-aluminate**	**PMMA**	**Ca-phosphates**
Biocompatibility	good	Poor	good
Elasticity, GPa	approx. 10	1-3	approx. 3
Compression strength, MPa	> 100	80-100	20-50
Hardness, Vickers	80-100	50-100	approx. 20
Mouldability	controllable	acceptable	too low
Thermal expansion coefficient, $^{o}C^{-1}$	approx. $9x10^{-6}$ $^{o}C^{-1}$	$>20x10^{-6}$ $^{o}C^{-1}$	-
Setting temperature, ^{o}C	38-40 ^{o}C	50-70 ^{o}C	Approx. 40 ^{o}C
Dimensional stability	< 0.5% expansion	1-2 % shrinkage	-
Long term stability, chemically	High stability	high	resorbable

For successful implantation in bone tissue early stabilization is of great importance. This includes both orthopaedic and dental implants. Even small gaps may lead to relative micro-motions between implant and the tissue, which increases the risk of implant loosening over time due to formation of zones of fibrous tissues at the implant-tissue interface. The effects of *in vivo* hydrating coatings systems on early anchoring were illustrated in Table VI above. In relation to the normal development of implant anchoring over time during the early stages of healing, the hydrating layers promote the early anchoring as a result of mass increase. The favourable removal torque value of paste augmented implants after two weeks may be interpreted as a result of the time needed for the paste to develop full mechanical strength.

In Table VIII is presented a comparison between a Ca-aluminate paste coating and a hydroxyapatite (HAP) coating – the totally dominating coating type.

Table VIII. Comparison between a Ca-aluminate paste coating and a hydroxyapatite coating

Feature	**Ca-aluminate paste coating (CA)**	**Hydroxyapatite coating (HAP)**	**Comments**
Application of layer	Paste or precursor	Sintered , PVD or CVD	CA layer develops *in vivo*
Stress towards substrate	Minimal stress	Stress level depending on the application technique	The *in vivo* (*in situ*) developed CA-hydrated layer formed by precipitation of nano crystals
Final contact zone to new tissue	Similar to that of HAP	Excellent -bioactivity	*In vivo* formed HA crystal in the surface of CA-paste coating towards tissue
Initial closure of gaps	partial	no	Dissolution /precipitation + mass increase close micro-voids to tissue wall.
Bioactivity	Chemical and biological integration	Biological intergration	The CA interacts with liquid body to form apatite (initially) and transformation of katoite to apatite in long-term tissue contacts
Early load capacity, Healing time	Short/ hours	weeks	loading ability after initial chemical reaction/mass increase of the CA-system, Ref [11]
Type of biomaterial	Chemically stable	Slowly resorbable	Stable materials for osteoporotic applications, while resorbable materials may be more suitable for young patients

CONCLUSIONS

Identified possible applications for CA-based materials are within vertebroplasty and odontology. An additional field of interest is as an implant coating material. Ca-aluminate based materials have a promising property profile with regard to handling, microstructure, and mechanical properties, biocompatibility and bioactivity. The hydrated CA materials are stable in bone tissue. Features contribute to the sealing of interfaces between the biomaterial and the surrounding tissue; first the general dissolution of Ca-aluminate and precipitation of nanosize hydrates in micro-voids between particles and between particles and tissue, and on all possible walls including tissue surfaces, second the mass increase due to water-up take from the surrounding tissue, and thirdly the zone formation including apatite formation upon which bioactive-induced formation of new tissue takes place. Thus both a pure chemical integration and a biologically induced integration contribute to sealing of biomaterial-tissue interfaces. The clotting tendency of the Ca-aluminate based biocement and the PMMA was much lower than for the calcium phosphate and calcium sulphate materials. The PMMA and the Ca-aluminate cement materials should not be expected to cause thrombotic complications in view of their very low tendency to induce activation of the coagulation system.

ACKNOWLEDGEMENT

The authors acknowledge the great contribution by cited colleagues at Doxa AB and the Angstrom Laboratory, Materials Science Department, Uppsala University. Financial support from Göran Gustafsson foundation for academic research is also acknowledged.

REFERENCES

[1]L. L. Hench, Biomaterials: a forecast for the future, *Biomaterials*, **19** (1998) 1419-1423.

[2]M. L. Roemhildt, T. D. McGee, S. D. Wagner, Novel calcium phosphate composite bone cement : strength and bonding properties, *Journal of Materials Science : Materials in Medicine*, **14** (2003) 137-141

[3]M. Nilsson, Injectable calcium sulphate and calcium phosphate bone substitutes, *Ph. D Thesis*, Lund (2003).

[4]L . Kraft, Calcium Aluminate Based Cement as Dental Restorative Materials, *Ph D Thesis*, Faculty of Science and Technology, Uppsala University, Sweden. 2002

[5]L. Hermansson, L. Kraft, H. Engqvist. Chemically Bonded Ceramics as Biomaterials, *Key Eng. Materials* **247** (2003) pp. 437-442. 2nd ISAC, 2002, Shanghai, China.

[6]N. Axen, T. Persson, K. Björklund, H. Engqvist, L. Hermansson. An Injectable Bone Void Filler Cement based on Ca-Aluminate . *Key Eng. Materials* **254-256** (2004) pp. 265- 268

[7] Patterson R J, Mayer D, Weaver L, Phaneuf M W. "H-Bar Lift-Out" and "Plan-View Lift-Out": Robust, Re-thinnable FIB- TEM preparation for ex-situ cross-sectional and plan-view FIB specimen preparation. *Submitted to Microscopy and Microanalysis.*

[8]H. Engqvist, J-E. Schultz-Walz, J. Loof, G. A. Botton, D. Maye , M. W. Pfaneuf, N-O.Ahnfelt , L. Hermansson "Chemical and Biological Integration of a Mouldable Bioactive Ceramic Material capable of forming Apatite in vivo in Teeth". *Biomaterials* **25** (2004) pp. 2781-2787

[9]H. Spengler, L.Hermansson and H. Engqvist, Evaluation of the Cohesiveness of Injectable Ca-aluminate based materials in water and simulated body fluid during curing, Extended abstract submitted for 16th Interdisciplinary Research Conference on Biomaterials

[10] N Axén N.-O. Ahnfelt, T. Persson, L. Hermansson, J. Sanchez, R. Larsson, A Comparative Evaluation of Orthopaedic cements in Human Whole Blood, Presented at Advanced Ceramics and Composites Conf , Cococa Beach 2005, to be published

[11]N.Axen, H Engqvist, J Loof, P Thomsen and L Hermansson, In vivo hydrating Calcium aluminate coatings for anchoring of metal implants in Bone, *Key Eng Mater.* **284-286** 831-834.

[12]J. Lööf, H.Engqvist, G. Gómez-Ortega, H. Spengler, N-O Ahnfelt and L.Hermansson, Mechanical Property Aspects of a Biomineral Based Dental Restorative System, Key Eng. Materials **Vols 284-286** , 741-7444 (2005)

[13]N. Axén, L.M. Bjursten, H. Engqvist, N-O Ahnfelt, L.Hermansson, Zone formation, presented at 9th ceramics: *Cells and tissue*, Sept 2004, to be published

[14] Y.Liu, J.Li, L. Sahlberg, L. Kraft, N-O Ahnfelt, L. Hermansson and J.Ekstrand, "Some Aspects of Biocompatibility and Chemical Stability of Calcium-Aluminate-Hydrate-Based Dental Restorative Material", *Presented at the IADR Chiba Conference*, Japan, June 2001

[15]T.C. Power and T.L. Brownyard, Studies of the physical properties of hardened Portland cement pastes, *J American Concrete Institute Proc* **43**, 492-498 (1946)

[16]J Lööf, H. Engqvist, L. Hermansson and N-O Ahnfelt, Mechanical Testing of Chemically Bonded Bioactive Ceramic Materials, Key Eng. Materials **Vols. 254-256** 51-54 (2004)

[17] H. Engqvist, E. Abrahamsson, J. Lööf and L.Hermansson , Microleakage of a Dental Restorative Material based on Biominerals, Presented at Advanced Ceramics and Composites Conf , Cococa Beach 2005, to be published

A THEORITICAL AND MATHEMATICAL BASIS TOWARDS DISPERSING NANOPARTICLES AND BIOLOGICAL AGENTS IN A NON POLAR SOLVENT FOR FABRICATING POROUS MATERIALS

Navin J. Manjooran and Gary R. Pickrell
Department of Materials Science and Engineering
Suite #302, 460 Turner Street
Virginia Polytechnic Institute and State University
Blacksburg, VA 24061

ABSTRACT

Porous materials are used in many different applications. These applications may require pores of different sizes, shapes and degree of connectivity. A novel method to fabricate porous materials has been described earlier where the pores in the material are produced by the death or the decomposition of the biological agent used. However to produce uniformity of the pores in the final structure, it is important to control the chemistry and rheological parameters during the fabrication process. Uniform dispersion of the nanoparticles around the biological agent is desired and the agglomeration of nanoparticles should be prevented. Dispersing nanoparticles with a suitable dispersant in the appropriate amount, while controlling the viscosity of the slurry are important factors and add an additional level of complexity when trying to fabricate materials with uniform porosity.

INTRODUCTION

Porous materials find applications as catalysts, sensors, electrodes, clothing, aerators and many other products [1-8]. The desired pore size, pore shape and degree of connectivity in a material varies depending on the material's final application. According to the International Union of Pure and Applied Chemistry (IUPAC), any solid material that contains cavities, interstices or channels may be regarded as a porous material [9]. Pores can be classified as open (have a continuous channel of communication with the external surface of the material) or closed (totally isolated from their neighbors) [9].

Recently, a novel methodology to fabricate porous materials has been discussed [1-5]. The novel porous material fabrication method has been used to produce porous ceramics and polymers using biological agents and nanoparticles. The biological agent used has been single celled fungi [2-4], bacteria [5] and viruses [10]. The size and shape of the pore formed depends on the size of the biological agent used and whether the biological agent produces a gas during processing. For generating nano-size scale porosity in materials, viruses can be used. Bacteria or single celled fungi can be used to generate micro-size scale porosity in materials.

PRELIMINARY RESULTS OF POROUS MATERIAL FABRICATED

Figure 1 is a scanning electron microscope (SEM) image of a silicon nitride porous material fabricated using 'pseudomonas aeruginosa' (bacteria) as discussed earlier [5]. In the image we can observe the large amount of porosity in the ceramic material. However, for many commercial applications, it is essential to control the distribution of pores in the material. By careful control of the rheology of the system and the mathematics involved with dispersing nanoparticles and the biological agents, we should be able to achieve uniform porosity in the material.

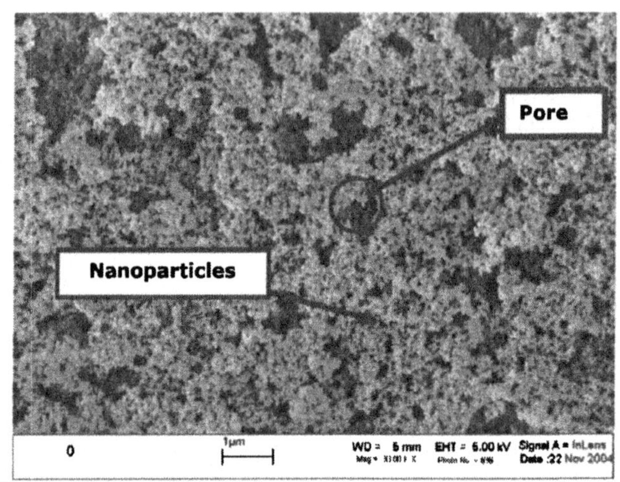

Figure 1. SEM image of porous Si_3N_4 fabricated using ‘pseudomonas aeruginosa’ (bacteria)

Figure 2 is an SEM image of a porous ceramic structure formed using type 2 Porcine circovirus (PCV2) (virus) by the novel technique as discussed earlier [10]. We need to understand the mathematics of dispersing nanoparticles and biological agents so as to achieve uniform porosity in the final structure fabricated.

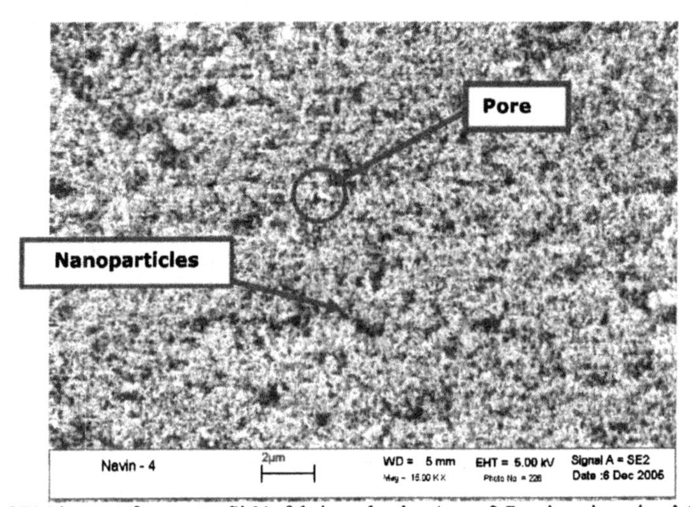

Figure 2. SEM image of a porous Si_3N_4 fabricated using ‘type 2 Porcine circovirus’ (virus)

In this paper we discuss the mathematics and the theoretical understanding required to disperse nanoparticles and biological agents in a media where only steric repulsion is present [non-polar media].

THEORETICAL AND MATHEMATICAL BASIS FOR DISPERSING NANOPARTICLES AND BIOLOGICAL AGENTS IN A NON-POLAR MEDIUM

The dispersion of nanoparticles is an area of considerable interest for various applications. Though a lot of work has been done with dispersing nanoparticles in a polar medium, dispersing nanoparticles in a non-polar system has not been sufficiently explored. The dispersing of nanoparticles with biological agents such as bacteria, fungi and viruses in a non-polar solvent is an interesting area which the authors believe has not been explored yet. An advantage of processing a well dispersed nanoparticle-biological agent system is that this may lead to fabricating materials with uniform porosity, by using a porous material production methodology described earlier [1-5].

To disperse the nanoparticles and the biological agent we need to understand the forces that are acting on them in the system, and balance the forces, so that the particle and biological agents do not aggregate. For simplifying the mathematical and theoretical understanding of the dispersing the nanoparticles and biological agents, let us make the following assumptions: [a] the forces acting between the particles (van der Waals) are attractive in nature, [b] the biological agent is spherical in shape [c] the biological agent is an extremophile [11, 12] that is stable in the non-polar solvent, [d] the particle and the biological agent have the same radius, and [e] the biological agent has properties similar to a particle.

The van der Waals forces are the main forces for coagulation of nanoparticles in a non-polar solvent, where only steric repulsion is present [13]. There are three contributions to the van der Waals interaction forces which are: [a] Keesom forces (interactions between permanent rotating dipoles), [b] Debye or induction interaction (between a permanent rotating dipole and an induced dipole), and [c] London or dispersion interaction (between two induced dipoles) [14]. Hence in the system, an appropriate balance [V_T] between the van der Waals attractive force [V_A] and steric repulsive forces [V_R] are needed to disperse the nanoparticles and biological agents in a non-polar solvent.

$$V_T = V_A + V_R \qquad \cdots [1]$$

This means that when the nanoparticle and the biological agent come close together, the van der Waals forces, which are attractive in nature, come into play and these are the major component for causing agglomeration in the system. However, if we can balance this attractive force [V_A] with a steric barrier [V_R], as shown in equation 1 [15, 16], a stable suspension should result. A schematic of a nanoparticle and a biological agent being separated by a steric barrier potential is shown in figure 3. When the nanoparticle and the biological agent covered with the adsorbed polymer layer approach each other due to the inter-penetration of their polymer layers, a repulsive force is produced (steric stabilization) [13]. When V_A=V_R, a balance between the repulsive and attractive forces will result and some optimum separation distance 'X' will be achieved. As shown in the figure, if we keep the nanoparticles and biological agents at a distance of 'X' from each other then the system is balanced with respect to the criteria mentioned above.

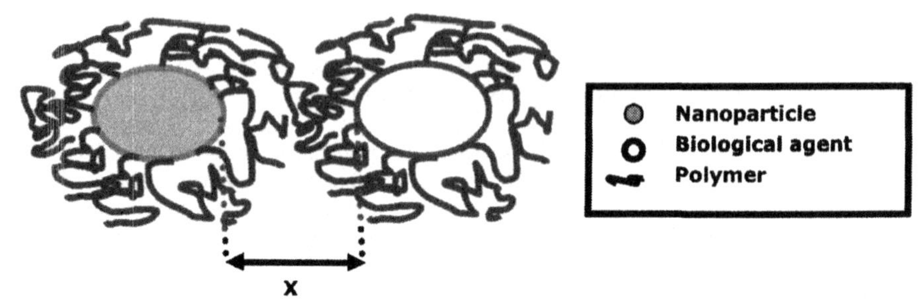

Figure 3. Sketch showing a nanoparticle and a biological agent being separated by a steric barrier potential

For steric stabilization of a colloidal suspension the following criteria need to be met: (a) the polymer should firmly adhere to the surface of the particle; (b) the stabilization moieties should have a good solvent condition,; (c) the adsorbed dispersant should be thick enough to overcome the attractive van der Waals forces; and (d) the polymer much cover the particle completely [13]. However, the attractive van der Waals forces are difficult to quantitatively measure with any surface force technique, since the forces increase by orders of magnitude close to the surface. This causes a jump in the measuring probe and no data can be collected. Hence it is important to determine theoretically the force balance discussed earlier.

In order to understand the origin of the van der Waals forces, we start with the equation for an ideal gas:

$$PV = nRT \quad \cdots [2]$$

where 'P' is the pressure, 'V' is the volume, 'n' is the number of moles, 'T' is the temperature and 'R' is the gas constant. The ideal gas will differ from the real gas in at least two ways. First in the calculation for the ideal gas, the molecules are treated as spherical points. Secondly, the forces of cohesion play an important role in a real gas which is not considered for the ideal gas. In the microscopic approach, the summation of pair wise interactions between the atoms in one body and the atoms in another body is used to derive an expression for the energy of attraction between them. Using this approach the van der Waals interaction energy [*W*] between two spherical particles may be expressed as shown in equation 3 [13].

$$W_{sph:sph} = \frac{A}{6}\left[\frac{2R^2}{H(4R+H)} + \frac{2R^2}{(2R+H)^2} + \ln\left(1 - \frac{4R^2}{(2R+H)^2}\right)\right] \quad \cdots [3]$$

where 'A' is the Hamaker's constant, 'R' is the particle radius and 'H' is the distance between the particle surfaces. Since the nanoparticle and the biological agent are assumed to be of the same radius (R), equation 3 can be used to calculate the van der Waals interaction energy between the nanoparticles and the biological agent. The Hamaker's constant for a particle (1)-

solvent (3)-biological agent (2) system (A_{132}) can be obtained from the Tabor-Winterton approximation [17, 18] as shown in equation 4, where 'k_B' is the Boltzmann's constant; 'T' is

$$A_{132}=\frac{3}{4}k_B T\frac{(\varepsilon_1-\varepsilon_3)(\varepsilon_2-\varepsilon_3)}{(\varepsilon_1+\varepsilon_3)(\varepsilon_2+\varepsilon_3)} \quad \text{... [4]}$$

$$+\frac{3h\nu_e}{8\sqrt{2}}\frac{(n_1^2-n_3^2)(n_2^2-n_3^2)}{(n_1^2+n_3^2)^{1/2}(n_2^2+n_3^2)^{1/2}((n_1^2+n_3^2)^{1/2}+(n_2^2+n_3^2)^{1/2})}$$

temperature in Kelvin; 'ε_1', 'ε_3', and 'ε_2' are the dielectric constants for the nanoparticle, the solvent and the biological agent respectively; 'n_1', 'n_3', *and* 'n_2' are the index of refraction for the nanoparticle, the solvent and biological agent respectively; 'h' is the Plank's constant and 'ν_e' is the plasma frequency (usually ~ $3*10^{15}$ Hz).

However, if the radius of the particles is greater than the separation distance between the particle surfaces, the expression in equation 3 can be reduced to the commonly used form represented as:

$$W_{sph:sph}=\frac{AR}{12H} \quad \text{... [5]}$$

where 'R' is the radius of the nanoparticle and 'A' is the Hamakar constant, which is defined as

$$A=\pi^2 C \rho_1 \rho_2 \quad \text{... [6]}$$

where 'ρ_1' and 'ρ_2' are the number of atoms per unit volume in the nanoparticles and 'C' is the coefficient of the atom-atom pair potential [19].

Using equation 5, we can predict the variation of van der Waals separation distance between the spherical nano-sized particles and the biological agent for a given system. An exponential curve will result when plotting equation 5, since the van der Waals interaction varies inversely with separation distance. From this curve, we can approximately determine the minimum distance (2r) to keep the particles separated which will be the point on the curve the attraction becomes significant. Expanding this knowledge we can determine the molecular weight of the dispersant selected for experimentation based on similarities in solubility parameters of the solvent and the dispersant [18]. Since the dispersant consists of repeated monomer units, calculating segment lengths between atoms by incorporating the radius of gyration equation [equation 7], with 'r' being half the minimum van der Waals distance calculated earlier, we can determine the appropriate order of magnitude for the molecular weight of the dispersant that can be used for experimentation [18, 20, 21].

$$r^2 = C\,N\,L^2 \quad \text{... [7]}$$

where 'L' is the segment length, 'C' is a constant [~5] and 'N' is the number of segments. This leads us to determine the amount (wt%) of dispersant needed for a specific volume percent of the nanoparticle suspension to completely cover the nanoparticles and biological agents in the slurry. To determine this we can use a ten volume percent (five volume percent each of the nanoparticle and the biological agent) solids loading slurry and obtain the variation of weight percent of dispersant versus viscosity for the slurry, using a rheometer. We should obtain a 'U' shaped curve as shown in figure 4. The viscosity of the slurry will be high at lower dispersant amounts due to insufficient polymer concentration in the slurry to completely cover the particles and the biological agent leading to bridging flocculation. The presence of an excess amount of dispersant also leads to an increase in the viscosity. The x-axis value on the lowest point [X] of this expected curve, figure 4, is the optimum weight percent of dispersant needed for a well dispersed slurry. Use of dispersant amounts less or greater than the optimum will lead to an increase in the suspension viscosity hampering the formation of a well dispersed slurry [21].

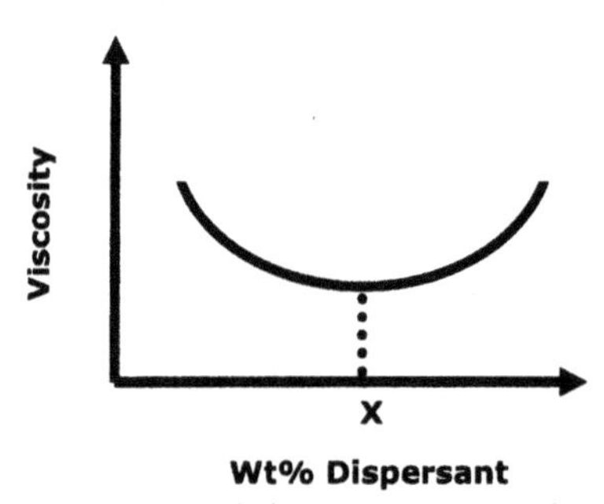

Figure 4. Sketch showing the variation of dispersant amount with viscosity

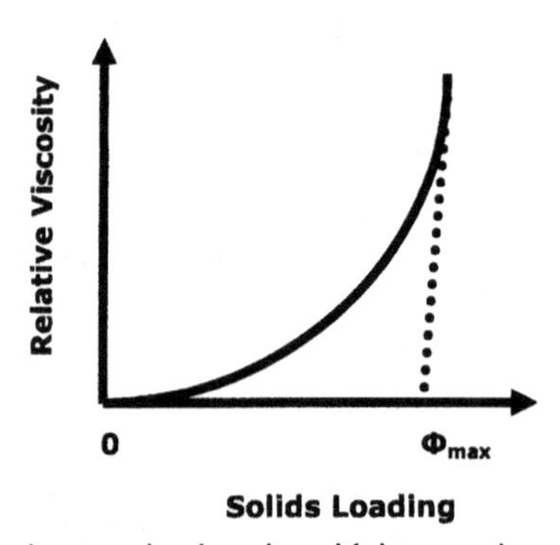

Figure 5. Sketch showing the increase in viscosity with increase in solids loading

Utilizing the optimum amount of dispersant and increasing the solids loading will result in an increase in viscosity of the slurry as shown in figure 5 as a greater shear stress will be needed for the same shear rate for the slurry [22]. Incorporating the Krieger-Dougherty fit [13, 17, 18]

shown in equation 8 to the data will result in determining the maximum solids loading for the slurry.

$$\eta_r = \left(1 - \frac{\phi}{\phi_m}\right)^{-[\eta]\phi_m} \quad \cdots \text{[8]}$$

where η_r is the relative viscosity , η is the intrinsic viscosity. Φ is the solids loading and Φ_m is the maximum solids loading.

The slurry obtained with the maximum solids loading can be cast into the appropriate shape and treated to elevated temperatures so as to produce porosity in the material using the porous material fabrication procedure described earlier [1-5].

LIMITATIONS

The theoretical and mathematical analysis described here can be applicable only to non-polar based systems. Biological agents are generally water based and hence understanding their dispersion with nanoparticles in a polar solvent will be important and is being currently investigated. Using a polar based system will make the porous material fabrication process easier, cheaper, and will generate lesser volatile gases during final processing.

CONCLUSIONS

The control of the rheological properties of the nanoparticle and biological agent suspension and the mathematics involved with dispersing the nanoparticles and biological agent is important to obtain a well dispersed suspension. The attractive van der Waals forces need to be balanced with a strong enough repulsive barrier so as to prevent agglomeration of nanoparticles and biological agents. The right amount of dispersant and its molecular weight can form the steric barrier potential needed to keep the nanoparticles and biological agents dispersed. From the Krieger-Dougherty equations we can determine the maximum solids loading for the system. The slurry with the maximum solids loading may be cast and treated to elevated temperatures to fabricate porous materials with uniform porosity.

REFERENCES

[1]Navin Manjooran, Erica Hartsell, and Gary Pickrell, "A Comparison of Porous Structural Ceramics and Porous Polymers Fabricated using Biological Agents and Nano Particles," in, *Materials Science and Technology Conference,* Pittsburg, PA, 195-202 (2005).

[2]Navin Manjooran and Gary Pickrell, "Complementary Fungus-Derived Micro-Porosity in Nano Materials," in, *107th Annual Meeting, Exposition and Technology Fair of the American Ceramic Society*, Baltimore, MD, 193-198 (2005).

[3]Navin J. Manjooran and Gary R. Pickrell, "Biologically Self-Assembled Micro and Nano-Structured Materials," *J. Materials Processing Technology,* Submitted (2005).

[4]Navin J. Manjooran and Gary R. Pickrell, "Biologically Self-Assembled Porous Polymers," *J. Materials Processing Technology,* 168, 225-229 (2005).

[5]Gary Pickrell and Navin Manjooran, "Biologically Derived Nano and Micro Porous Material," in, *107th Annual Meeting, Exposition and Technology Fair of the American Ceramic Society*, Baltimore, MD, 229-237 (2005).

[6]K. Ishizaki, S. Komarneni, and M. Nanko, *Porous Materials: Process Technology and Applications*, (1998).
[7]G. Q. Lu and X. S. Zhao, *Nanoporous Materials: Science and Engineering*, (2004).
[8]T. J. Pinnavaia and M. F. Thorpe, *Access in Nanoporous Materials*, (1995).
[9]J. Rouquerol, D. Avnir, C. W. Fairbridge, D. H. Evereit, J. H. Haynes, N. Pernicone, J. D. F. Ramsay, K. S. W. Sing, and K. K. Unger, "International Union of Pure and Applied Chemistry: Recommendations for the Characterization of Porous Solids," *Pure and Applied Chemistry* 66, 8, 1739-1758 (1994).
[10]Navin J. Manjooran, Nicole E. McKeown, Diane Fields, Gary R. Pickrell, and X. J. Meng, "A Novel Biologically Derived Methodology Towards Fabrication of Nanoporous Ceramics Using Porcine Circovirus (PCV2) and Nanoparticles," in, *30th International Conference and Exposition on Advanced Ceramics and Composites*, Cocoa Beach, Florida, Submitted, (2006).
[11]S. Isken and Jan A. M. de Bont, "Bacteria tolerant to organic solvents," *Extremophiles* 2, 3, 229 - 238 (1998).
[12]R. Margesin and F. Schinner, "Biodegradation and bioremediation of hydrocarbons in extreme environments," *Applied Microbiology and Biotechnology* 56, 5-6, 650-663 (2001).
[13]Wolfgang M. Sigmund, Nelson S. Bell and Lennart Bergstrom, "Novel Powder-Processing Methods for Advanced Ceramics," *Journal of American Ceramic Society* 83, 7, 1557-1574 (2000).
[14]Seung-woo Lee and Wolfgang M. Sigmund, "Repulsive van der Waals Forces for Silica and Alumina," *Journal of Colloid and Interface Science* 243, 365-369 (2001).
[15]J. L. Ortega-Vinuesa, A. Martin-Rodriguez, and R. Hidalgo-Alvarez, "Colloidal Stability of Polymer Colloids with Different Interfacial Properties: Mechanisms," *Journal of Colloid and Interface Science* 184, 259-267 (1996).
[16]M. S. Romero-Cano, A. Martin-Rodriguez, G. Chauveteau and F. J. de las Nieves, "Colloidal Stabilization of Polystyrene Particles by Adsorption of Nonionic Surfactant," *Jouranal of Colloid and Interface Science,* 198, 273-281 (1998).
[17]Lennart Bergstrom, "Hamakar Constants of Inorganic Materials," *Advances in Colloid and Interface Science,* 70, 125-169 (1997).
[18]Navin Manjooran, "Development of an Alpha Silicon Carbide Based Liquid Toner for Electro-Photographic Solid Freeform Fabrication" (Electronic, University of Florida, (2003).
[19]Seung-woo Lee and Wolfgang M. Sigmund, "AFM Study of Repulsive van der Waals Forces Between Teflon AF^{TM} Thin Films and Silica or Alumina," *Colloids and Surfaces A: Physiochemical and Engineering Aspects* 204, 43-50 (2004).
[20]M. N. Rahman, *Ceramic Processing and Sintering*, (1995).
[21]Navin Manjooran, Ashok Kumar and Wolfgang Sigmund, "Development of a Liquid Toner for Electro-photographic Solid Freeform Fabrication," *J. European Ceramic Society,* In Press (2005).
[22]Dennis R. Dinger, *Rheology for Ceramists*, (2002).

PREPARATION OF HYDROXYAPATITE AND CALCIUM PHOSPHATE BIOCERAMIC MATERIALS FROM THE AQUEOUS SOLUTION AT ROOM TEMPERATURE

Jia-Hui Liao, Yu-Chen Chang, and Tzer-Shin Sheu
Department of Materials Science and Engineering
I-Shou University, Kaohsiung, Taiwan

ABSTRACT

A chemical reaction method was used to study the formation mechanism of hydroxyapatite on the surfaces of $Ca_3\,(PO_4)_3$, $Ca_3\,(PO_4)_3/Al_2O_3$, and $Ca_3\,(PO_4)_3/ZrO_2$ ceramic substrates, when these ceramic substrates were chemically reacted with Ca^{2+}, PO_4^{3-} and OH^- ions in the different aqueous solutions. With a dipping coating or soaking technique, brushite (dicalcium phosphate dihydrate, $CaHPO_4 \cdot 2H_2O$) was formed on the surface of ceramic substrate in the phosphate acid solution at pH=2.5 . After a subsequent solution treatment in a 2.5M of NaOH aqueous solution at 80°C for 1h, brushite coating was converted into hydroxyapatite (HA). Except for the phase change, the morphologies of surface coatings were also progressively changed from the plate-like structure to the network structure as the pH value increased. In the $Ca_3\,(PO_4)_3/Al_2O_3$ and $Ca_3\,(PO_4)_3/ZrO_2$ ceramic substrates, HA was much easier to form on these substrate surfaces after an alkaline solution treatment, if the volume ratio of $Ca_3\,(PO_4)_3/Al_2O_3$ or $Ca_3\,(PO4)_2/ZrO_2$ in the ceramic substrate was 7/3. As to the effect of basicity, HA was much easier to form on the ceramic substrates at pH=12-14 after a subsequent NaOH-containing solution treatment.

INTRODUCTION

Calcium phosphate ceramics have been widely used in the clinical applications because of their biocompatibility and osteoconduction.[1-2] However, most calcium phosphate ceramics do not have excellent mechanical properties to sustain a severe loading environment, such a dental application in the oral environment. Therefore, other materials or ceramics, such as alumina (α-Al_2O_3) and yttria-stabilized zirconia (YSZ), are chosen as the substrate for strengthening calcium phosphate coating, because they have excellent mechanical strength and wear resistance behaviors. Currently, several coating techniques like electrophoretic method, electrochemical deposition, plasma spray, and sol gel coating methods have been used to obtain a good HA coating on the commercial implant materials.[3-8]

In the past decade, the hydroxyapatite coating has been effectively used to improve the bonding strength between the metallic implant and bone. However, most of metallic implants often release some detrimental ions to cause some unexpected illnesses or symptoms after implantation in human body. In this study, bio-inert or bioactive active ceramic composite materials were chosen as the substrates, and a chemical solution method was used to let hydroxyapatite formed on the surface of these substrates. This chemical solution method was very easy to operate, and it was a low-cost process. Ceramic substrates were firstly soaked in the phosphate acid solution, and then were further treated in the alkaline solution to obtain a HA coating or deposition on the surface of ceramic substrates. More interestingly, through such a simple dissolution-precipitation mechanism, it is expected that the bone-like apatite layer can be mimetically formed on the different ceramic substrates. Therefore, in this study, processing factors including pH, temperature, and the concentrations of acid and alkali solutions are critically controlled to observe their effects in the formation and microstructure of HA coating.

EXPERIMENTAL PROCEDURES

Starting chemicals or powders were HA (37% pure of calcium, Strem Chemicals, America), $Ca_3(PO_4)_2$ (96% pure, Fluka, Germany), Al_2O_3 (99% pure, Osaka, Japan), and ZrO_2+5wt%Y_2O_3 (99.95% pure, Cerac, America). A batch of $Ca_3(PO_4)_2$, $Ca_3(PO_4)_2/Al_2O_3$, and $Ca_3(PO_4)_2/ZrO_2$ well-mixed powders was separately cold-pressed to form a green powder compact under a uni-axial pressure of 50 MPa. The cold-pressed green powder pellets were then sintered at 1400°C for 0.5h in air, with a heating rate of 10°C/min, to obtain a porous or densified ceramic substrate. Sintered $Ca_3(PO_4)_3$, $Ca_3(PO_4)_2/Al_2O_3$, and $Ca_3(PO_4)_2/ZrO_2$ ceramic composites were polished, cleaned, and dried to remove any surface contaminants. These polished ceramic substrates were then soaked in the 1 M H_3PO_4 aqueous solution at T=25-80°C under pH=0.7~11 to observe phase existence and microstructure on the substrate surface.

X-ray diffraction (XRD) method was used to determine phase existence on the ceramic substrate surface. Microstructure observation was conducted by a scanning electron microscope (SEM). Reflectance absorption infrared spectroscopy (RAIR) was used to determine any new bonding formed on the ceramic substrate surface.

RESULTS AND DISCUSSION

(A) Phase existence in the sintered ceramic substrates

Phase existence in the different sintered substrates is listed in Table 1. Sintered $Ca_3(PO_4)_3$ ceramic substrate, represented by the symbol "Ca_3P", contained β-TCP as the sintered specimen was cooled to room temperature. Sintered Ca_3P/ZrO_2 substrate contained m-ZrO_2 (monoclinic zirconia) and β-TCP.

Table1. Phase existence in the different sintered ceramic substrates

Substrate	Nominal Composition	Phase Existence*	Particle Size (μm)
HA	$Ca_{10}(PO_4)_6(OH)_2$	HA, TTCP	0.13
Ca_3P	$Ca_3(PO_4)_2$	β-TCP	5
Ca_3P/Al_2O_3	$Al_2O_3 + Ca_3(PO_4)_2$	α-Al_2O_3, β-TCP	5/50
Ca_3P/ZrO_2	$ZrO_2 + Ca_3(PO_4)_2$	m-ZrO_2, β-TCP	5/26

*HA, $Ca_{10}(PO_4)_6(OH)_2$; TTCP, $Ca_4P_2O_9$; β-TCP, $Ca_3(PO_4)_2$; m-ZrO_2, monoclinic ZrO_2

(B) After solution treatments in the phosphate acid and NaOH solutions

X-ray diffraction patterns for a solution-treated Ca_3P substrate are shown in Fig 1(a). This solution-treated Ca_3P substrate was soaked in the phosphate acid solution at pH= 2.5 with a rich Ca^{2+} environment for 1h. It indicates a crystalline brushite coating exists on the $Ca_3(PO_4)_2$ substrate surface. After a subsequent solution treatment in the 2.5M of NaOH solution, the brushite surface coating was converted into a single crystalline hydroxyapatite phase, which can be seen from X-ray diffraction patterns shown in Fig. 1(b). The chemical reaction for a subsequent alkaline treatment is suggested to be $10\ CaHPO_4{\bullet}2H_2O + 2\ OH^- \rightarrow Ca_{10}(PO_4)_6(OH)_2 + 20\ H_2O + 10\ H^+ + 4\ PO_4^{3-}$. The same phenomenon of the phase conversion from brushite to hydroxyapatite also occurred in the solution-treated Ca_3P/ZrO_2 and Ca_3P/Al_2O_3 substrates, as shown in Figs. 2 & 3. However, only a small amount of the brushite-to-hydroxyapatite conversion was found in the Ca_3P/Al_2O_3 substrate after a subsequent solution treatment in the 2.5M of NaOH solution. This was because brushite was not formed extensively after a first solution treatment in the phosphate acid solution.

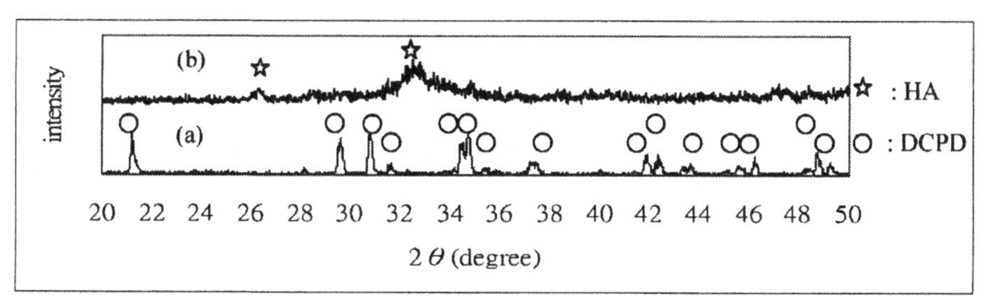

Fig. 1 X-ray diffraction patterns for the Ca_3P substrate (a) after a solution treatment in the phosphate acid solution at pH=2.5, and (b) after a subsequent treatment in the 2.5M of NaOH solution for 1h.

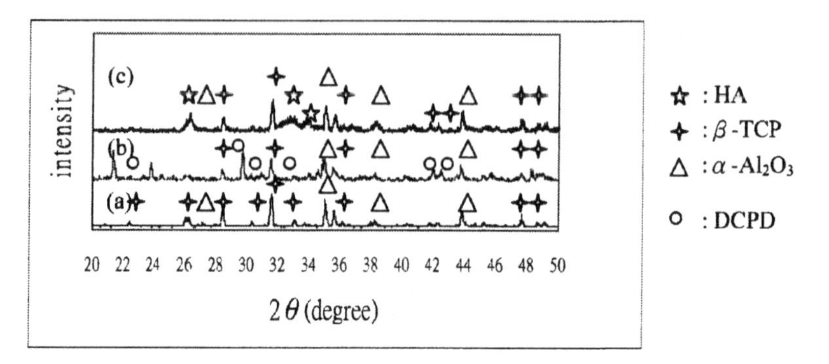

Fig. 2 X-ray diffraction patterns for the Ca_3P/Al_2O_3 substrate (a) before a solution treatment, (b) after a solution treatment in the phosphate solution at pH=2.5, and (c) after a subsequent solution treatment in the 2.5M of NaOH solution.

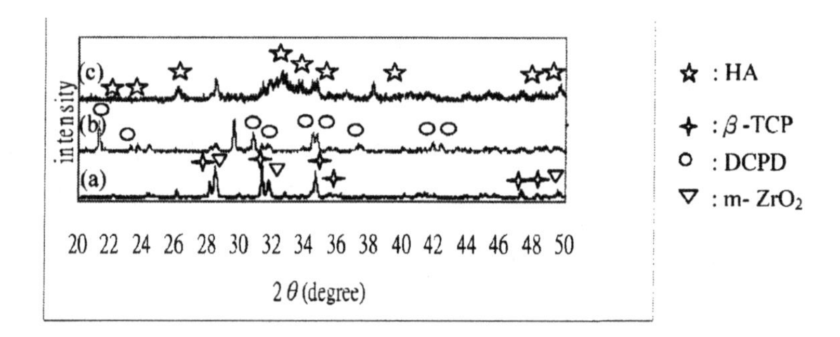

Fig. 3 X-ray diffraction patterns for the Ca_3P/ZrO_2 substrate (a) before a solution treatment, (b) after a solution treatment in the phosphate solution at pH=2.5, and (c) after a subsequent solution treatment in the 2.5M of NaOH solution.

SEM micrographs of surface morphologies on different ceramic substrates after a solution treatment in the phosphate acid solution and a subsequent solution treatment in the 2.5M of NaOH solution are shown in Fig. 4. Plate-like brushite crystals were observed on the different substrate surfaces after a solution treatment in the phosphate solutions, as shown in Figs. 4(a)-(c). However, brushite crystals were distributed very uniformly on the Ca_3P and Ca_3P/ZrO_2 substrate surfaces, as shown in Figs. 4(a) & 4(c), but very limitedly and locally on the Ca_3P/Al_2O_3

substrate surface, as shown in Fig. 4(b). After a subsequent solution treatment in the 2.5M of NaOH solution, a fine network-like hydroxyapatite crystals were formed on the ceramic substrate surfaces, as shown in Figs. 4(d)-(e).

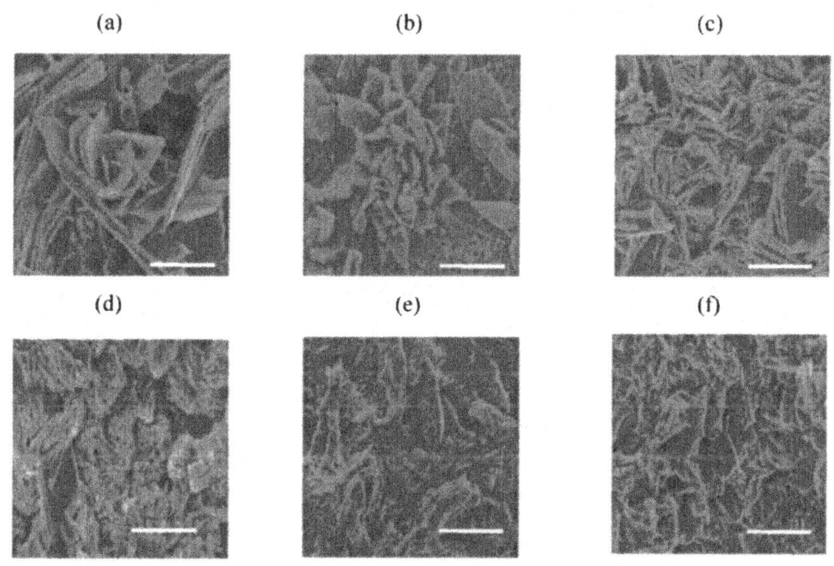

Fig. 4 Surface morphologies for the substrates (a) Ca_3P, (b) Ca_3P/Al_2O_3, and (c) Ca_3P/ZrO_2 after a solution treatment in the phosphate acid, and for the substrates (a) Ca_3P, (b) Ca_3P/Al_2O_3, and (c) Ca_3P/ZrO_2 after a subsequent solution treatment in the 2.5M of NaOH aqueous solution. Each scale bar is for 30 μ m.

(C) Infrared spectroscopy

Reflectance Absorption Infrared Spectroscopy (RAIR) spectra after a subsequent solution treatment in the 2.5M of NaOH solution on the different ceramic substrates are shown in Fig. 5. The RAIR spectra shown on Fig. 5 indicate that all ceramic substrate surfaces contain several absorption peaks at the wavenumber of 3400~3600 plus 600~700 cm^{-1} for the OH bond, and 1000~1200 plus 500~600 cm^{-1} for the PO bond, which are the characteristic ionic bonds for hydroxyapatite ceramic. Again, the formation of hydroxyapatite coating is further confirmed from the determination of RAIR spectra.

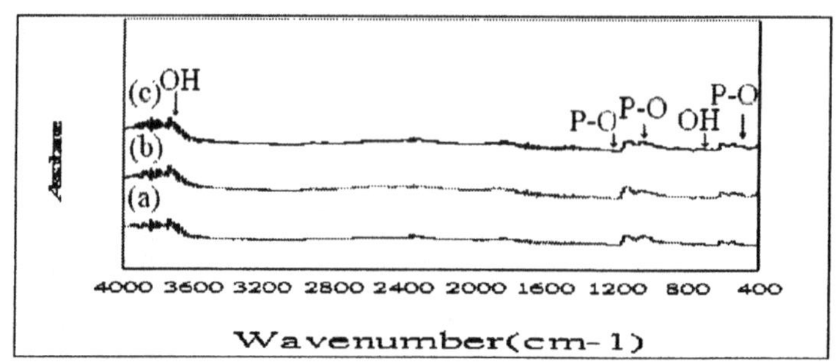

Fig. 5 RAIR spectra for ceramic substrates (a) Ca_3P, (b) $Ca_3P/3Al_2O_3$, and (c) Ca_3P/ZrO_2 after a subsequent solution treatment in the 2.5M of NaOH solution.

(D) A subsequent solution treatment in different alkali solutions

Different alkali solutions including KOH, NaOH, and NH_4OH were used to do a subsequent treatment for the phosphate-acid-treated ceramic substrates. X-ray diffraction patterns for the Ca_3P substrate after a subsequent solution treatment in different alkali solutions are shown in Fig. 6. These alkali-solution-treated Ca_3P substrate surfaces all contained hydroxyapatite and with much broader characteristic HA diffraction peaks. From the broadening X-ray diffraction peaks, it demonstrates that the hydroxyapatite coating contains a very fine crystalline structure.

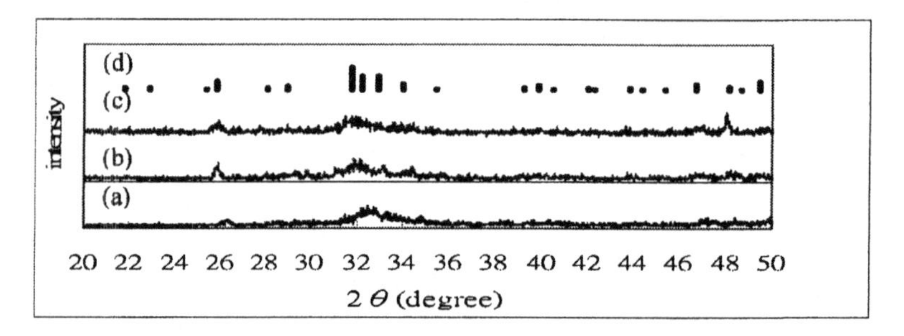

Fig. 6 X-ray diffraction patterns for the Ca_3P substrate surfaces after the alkali solution (a) NaOH, (b) KOH, and (c) NH_4OH treatment, and (d) HA diffraction patterns from JCPDS 09-0432.

CONCLUSIONS

A hydroxyapatite coating was prepared from a chemical solution method, in which ceramic substrates $Ca_3(PO_4)_2$, $Ca_3(PO_4)_2/Al_2O_3$, and $Ca_3(PO_4)_2/ZrO_2$ were firstly soaked in the phosphate acid solution, and were subsequently treated in the NaOH, KOH, or NH_4OH solution at room temperature. After a solution treatment in the phosphate acid solution, brushite was formed uniformly on the $Ca_3(PO_4)_2$ and $Ca_3(PO_4)_2/ZrO_2$ substrate surfaces, but locally on the $Ca_3(PO_4)_2/Al_2O_3$ substrate surface. This brushite coating contained a plate-like crystalline structure.

After a subsequent solution treatment in the 2.5M of NaOH solution, plate-like brushite crystals were converted into network-like hydroxyapatite crystals. The content of hydroxyapatite crystals on the ceramic surfaces was lower on the $Ca_3(PO_4)_2/Al_2O_3$ substrate surface.

With a volume ratio of $Ca_3(PO_4)_2/ ZrO_2=7/3$ or $Ca_3(PO_4)_2/ Al_2O_3=7/3$, hydroxyapatite was much easier to form in a subsequent alkaline solution treatment. Except for that, hydroxyapatite could be formed in the NH_4OH solution and different alkaline solutions like KOH and NaOH. From X-ray diffraction patterns and SEM micrographs, the hydroxyapatite coating contained a very fine polycrystalline HA particles.

REFERENCE

[1]M.V. Regi, J.M.G. Calbet, "Calcium phosphates as substitution of bone tissues", *Prog. Solid State Chem.*, **32**, 1–31 (2004).

[2]C. Nicolazo, H. Gautier, M.J. Brandao, G. Daculsi, and C. Merle, "Compatibility study of calcium phosphate biomaterials", *Biomaterials*, **24**, 255–262(2003).

[3]Y.Z. Yang, K.H. Kimc, and J.L Onga, "A review on calcium phosphate coatings produced using a sputtering process—an alternative to plasma spraying", *Biomaterials*, **26**, 327–337(2005).

[4]S. V. Dorozhkin, "A Review on the Dissolution Models of Calcium Apatites", *Prog. Cryst. Growth Charact. Mater.*, **44**, 45-61(2002).

[5]M.C. Kuo, S.K. Yen, "The process of electrochemical deposited hydroxyapatite coatings on biomedical titanium at room temperature", *Mater. Sci. Eng.*, **20**, 153–160 (2002).

[6]Q.L. Feng, F.Z. Cui, H. Wang, T.N. Kim, and J.O. Kim, "Influence of solution conditions on deposition of calcium phosphate on titanium by NaOH-treatment ", *J. Cryst. Growth*, **210**, 735-740(2000).

[7]M. H. Prado, D. Silvaa, J. H. C. Limac, G.A. Soares, C.N. Eliasd, M.C. de Andradee, S.M. Bestf, and I.R. Gibson, "Transformation of monetite to hydroxyapatite in bioactive coatings on titanium", *Surface and Coatings Technology*, **137**, 270-276(2001).

[8]C. S. Liu, Y. Huang, W. Shen, and J.G. Cui, "Kinetics of hydroxyapatite precipitation at pH 10 to 11", *Biomaterials*, **22**, 301-306 (2001).

HYDROXYAPATITE COATINGS PRODUCED BY PLASMA SPRAYING OF ORGANIC BASED SOLUTION PRECURSOR

E. Garcia, Z. B. Zhang, T. W. Coyle
Centre of Advanced Coating Technologies, University of Toronto.
184 College St.
Toronto, ON, M5S 3E4

L. Gan, R. Pilliar
Faculty of Dentistry, University of Toronto
124 Edward St.
Toronto, Ontario, M5G 1G6

ABSTRACT

Liquid precursors and suspensions are being explored as new feedstock for air plasma spray (APS) due to the low cost of production and submicron/nanostructured features that are found in the coatings produced in this way. The goal of this work is to study the suitability of a liquid precursor of hydroxyapatite (HA) as feedstock for the APS coating technique. The precursor is an organic sol-gel solution of calcium nitrate tetrahydrate and triethyl phosphite that was employed in a previous work to produce thin films by a dipping technique. The liquid precursor is injected in the plasma plume by a pressurized system and the coatings are deposited on Ti6Al4V substrates. The hydroxyapatite coatings so formed are porous and are characterized by microstructure features typically found in solution precursor plasma spray processes; a combination of melted and unmelted deposits and small hollow spheres. The analysis of the coatings by transmission electron microscopy reveals submicron/nanocrystalline features forming those deposits. X ray diffraction analysis shows that hydroxyapatite is the main crystalline phase present in the coatings. Small amount of other crystalline phases product of the decomposition of the hydroxyapatite, are also found due to the high temperature of the substrates during deposition.

INTRODUCTION

Plasma sprayed hydroxyapatite (HA) has been used extensively as surface coatings on metallic implants since the mid-1980's[1,2] . These biomedical implants are generally made of biocompatible stainless steels, CoCr alloys, or Ti alloys. The coatings have been shown to promote bone fixation and osteoconductivity. However, there are some concerns related to the formation of large particle debris during the resorption of the coating[3]. To overcome this situation coatings formed by smaller features are required. Submicron/nanocrystalline HA coatings may improve the resorption of the coating in the body, avoiding the irritant effect of large particles which may be seen in current thermal sprayed HA coatings.

A modification of the conventional plasma spray technique, solution precursor plasma spray (SPPS), has been reported to produce submicron/nanocrystalline structured coatings from relatively inexpensive precursors[4]. In this process, a solution of the coating precursors is atomized and injected into a direct current (DC) plasma jet. The precursor droplets undergo rapid evaporation and breakup in the plasma. This is followed by precipitation and pyrolysis. The operative mechanisms in SPPS were found to be critically dependent on the trajectories of the precursor droplets in the plasma jet[5]. The typical features vary from undecomposed precursor droplets to adherent deposits. The adherent deposits are obtained when the precursors are injected to the hot area of the plasma where they melt and solidify upon arriving at the substrate.

The solution precursors used in this work were developed previously for a sol-gel deposition process. Sol-gel Ca-P/HA thin coatings formed by either a so-called inorganic or organic route have been reported in the literature[6]. In general, these liquids consist of a colloidal suspension of inorganic hydroxide particles or metal alkoxide or other organic precursors.

The purpose of the present investigation was to check the suitability of a liquid organic precursor to produce Ca-P coatings using the solution plasma spray deposition technique.

MATERIALS AND EXPERIMENTAL PROCEDURE

Solution Precursor

The organic route Ca-P sol-gel solution was prepared according the methods described by Liu et al.[7]. Calcium nitrate tetrahydrate (Sigma-Aldrich Inc., St. Louis, MO) and triethyl phosphite (Sigma-Aldrich Inc., St. Louis, MO) were the calcium and phosphorous precursors, respectively. Triethyl phosphite was first hydrolyzed in a small amount of distilled deionized H_2O. The mixture was sealed and stirred vigorously for 24 h to allow complete hydrolysis of triethyl phosphite. Calcium nitrate tetrahydrate dissolved in absolute anhydrous ethanol was added dropwise to obtain a Ca/P ratio of 1.67. The resulting clear solution was sealed and aged at 40 °C in a water bath for 4 days. It was not possible to measure the particle size by laser scattering size distribution analyzer. The particle size reported by Liu et al.[7] after evaporating the solvent and heating the solid residue at 350 °C was 0.120 μm. The concentration of solids was 12 wt.%, determined by measuring the mass of the residue formed after evaporating the solvent and heating to500 °C.

Sol Plasma Spraying

The sol-gel solution was sprayed using a DC arc plasma SG-100 gun (Miller Thermal Inc., now Praxair Surface Technology, Indianapolis, IN). The main plasma gas used was Argon at a flow rate of 50 slpm. A flow of 2 slpm of H_2 was used as a secondary plasma gas. The solution was injected from a pressurized vessel using a home made atomizing nozzle fixed at 3 mm from the exit of the gun and tilted 20° up towards the gun to ensure the proper injection of the precursor into the plasma core. The atomizing pressure was 35 psi and the feeding rate was adjusted to 30 ml/min in all cases. Three cycles of 10 passes each were used to obtain the coatings. Before the coating deposition, a deposit footprint was obtained by briefly exposing a mirror polished stainless steel substrate while holding the gun stationary..

The substrates were made from mill-annealed Ti6Al4V alloy (12.5 mm in diameter 3 mm thick). Prior to deposition they were sand blasted with alumina grit. The substrates were fixed at 6 cm from the torch and during spraying were air cooled to avoid overheating.

Figure 1 is a schematic of the solution precursor plasma spray system.

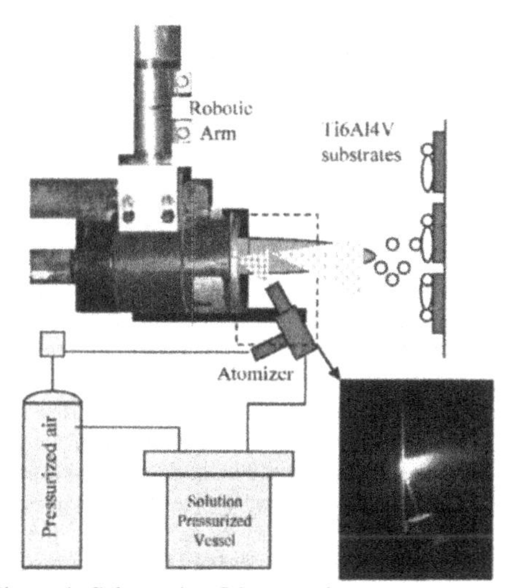

Figure 1: Schematic of the spraying system set up.

Characterization of the Coatings

Coated substrates were cut with a diamond saw (Clemex, Brillian 221, Clemex Technologies Inc., Longueuil, QC, Canada) after protecting the coating by covering with a low viscosity resin. The cut samples were mounted using the same low viscosity resin. After mounting the specimens were polished to a 1200 grit finish on a semi-automatic polisher (Imptech DPS 2000, Boksburg, South Africa) using alcohol as the polishing media. The mounted samples were gold or graphite coated to make the mounting resins conductive. In other cases the coatings were scrapped from the substrates, dispersed in ethanol and deposited on copper grids to analyze them by transmission electron microscopy.

The thickness, morphology and microstructure of the coatings and the features obtained in the footprint test were analyzed by scanning electron microscopy (SEM, Hitachi S-2500, Hitachinaka-shi, Ibaragi, Japan), field emission scanning electron microscopy (FE-SEM, Hitachi S-4500. Hitachinaka-shi, Ibaragi, Japan), and high resolution scanning transmission microscopy (STEM, Hitachi HD-2000, Hitachinaka-shi, Ibaragi, Japan). Energy Dispersive X-ray analysis (EDX) attached to the SEM was employed to determine qualitatively the chemical composition of the coatings.

Crystalline phases present in the original sol dried at 100 °C, calcined at 500 °C and in the as-sprayed coatings were determined by X ray diffraction analysis (CN2651A1, Rigaku Co. The Woodlands, TX) using Cu Kα radiation. Scan parameters included a scan range 2θ=20-45°, a step size of 0.02° and a step time of 1.55 s.

RESULTS AND DISCUSSION

Microstructure of the Coatings

The solution was sprayed using the spraying parameters mentioned in the previous section. It was easy to observe a long plasma plume attributed to the decomposition of the ethanol present in the organic precursor. That long flame impinged on the substrates during the spraying and, although the substrates were air cooled, in some cases were heated to red hot, which means temperatures between 700- 800 °C.

In Fig. 2 are shown scanning electron micrographs illustrating different features found in the footprint test. Figures 2a and 2b were obtained in the outer area of the plasma footprint (~ 2 cm from the center). This area is characterized by the presence of the features that come from the precursor subjected to the lower plasma temperatures. In this case, the feature is a gel drop that has been dried upon its arrival at the substrate. Together with the dried gel splat can be observed several spheres. The presence of these spheres is characteristic of the solution precursor plasma spray method and could be related to small droplets injected in an area of medium temperatures of the plasma flame. In Fig. 2c and 2d are shown typical splats corresponding to the center part of the footprint, i.e. the area where the precursors subjected to the highest temperatures are deposited. Again, small spheres are found together with the splats formed from molten material. In some cases these spheres are hollow. The coatings will be a combination of the features found in the different areas of the footprint.

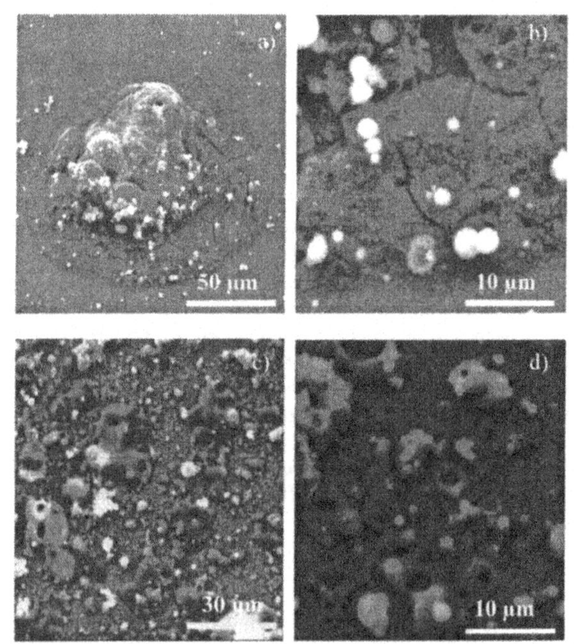

Figure 2: SEM micrographs of the footprint test: a) and b) outer area and c) and d) center of the footprint.

In Fig. 3 are shown the top view of the coating and a detail of the same area at higher magnification obtained by FE-SEM. It can be observed that the coating has very porous open microstructure at the uppermost surface. The high porosity of the coatings could enhance the bone attachment to the implant. Image b) of Fig. 3 show how these coatings are formed by submicron spheres joined together to form the coating. The submicron size of these features may avoid the problems related with the resorption of the coating during the bone ingrowth, found in traditional plasma sprayed HA coatings[3,8].

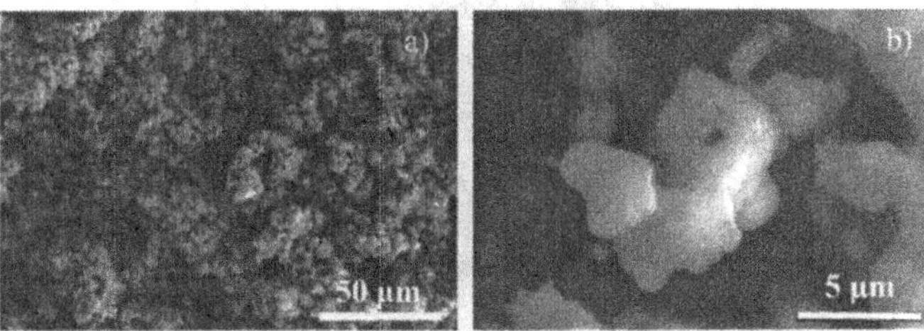

Figure 3: Top view FEM micrographs of the HA coating.

The overall cohesion of the coating however was not very good although some spheres were strongly attached to others as a result of partial melting of their surfaces. Transmission electron microscopy was performed to study in more detail how those particles are joined together to form the coating. In Fig. 4 STEM micrographs of a sample scrapped from the original coating are shown. Figure 4a was obtained in the scanning mode of the STEM and Fig. 4b in the transmission mode. The particle can be seen to consist of an aggregation of much finer particles, perhaps the original sol particles.

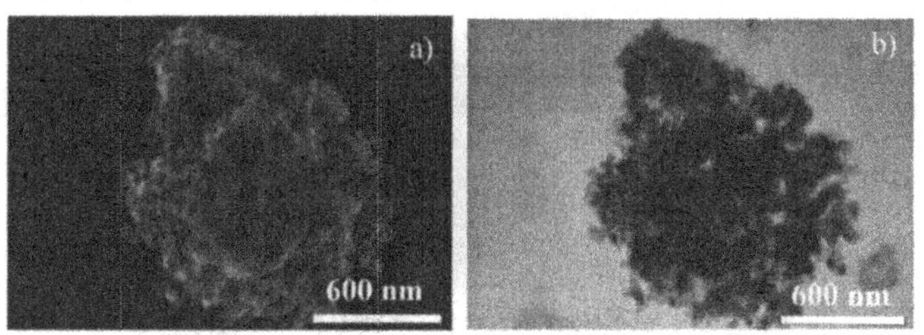

Figure 4: STEM micrographs of samples scrapped from the coatings.

In Fig. 5 are shown backscattered electron SEM micrographs of polished cross section of the as sprayed coating. The image shows a porous coating with a thickness of approximately 50 μm. In this cross section can be seen a small number of melted particles, hollow spheres with a dense skin and unmelted porous aggregates of nanoparticles, which act as the "cement" that join the hollow spheres.

In Fig. 5 can also be seen a dense layer 3-5 μm thick between the porous HA coating and the Ti6Al4V substrate. Qualitative EDX analyses revealed a composition for that layer containing Ti, Al, V, Ca and P. This layer is the result of the reaction between HA and the titanium substrate under the high temperature and oxidizing conditions[9] during the spraying. The presence of this layer can be explained by the high temperature of the substrate during deposition. As mentioned before, the presence of ethanol in the composition of the organic precursor produced a long flame that increased the temperature of the substrates. Although the presence of a reaction layer has been reported as a possible cause of enhanced interfacial shear strength between the substrate and the coating[10], it must be very thin to perform in this way. EDX analyses did not reveal the presence of Cu or W-Th elements coming from the cathode or anode of the plasma gun. Further analysis with a more sensitive technique will be done.

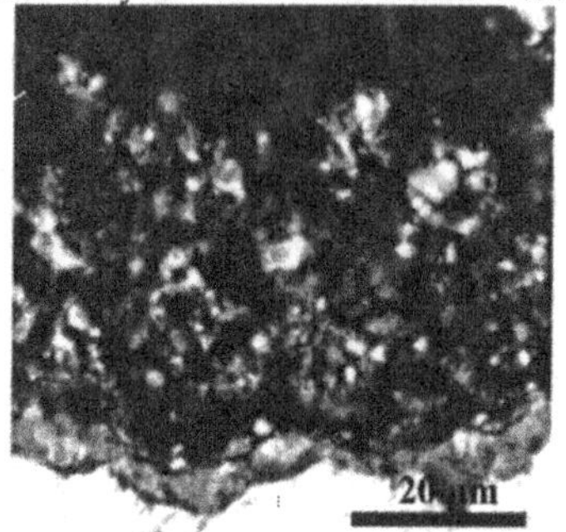

Figure 5: Backscattered electron SEM micrograph of polished cross section of the coating.

The deposition mechanism of the coatings is evident after analyzing the previous figures and the features found in the footprint test. The three different constituents of the coating: melted features, hollow spheres and the "cement" formed by aggregates of nanoparticles are related to the penetration of the precursor into the plasma plume[5]. Melted particles are related to precursor droplets injected into the hottest area. The hollow spheres come from an area with a lower temperature, and the nanostructured aggregates are droplets from the colder fringe of the plasma plume that arrive at the substrate in the form of wet splats and are dried on the substrate by the heat generated by the plasma flame.

The cohesion of the nanostructured aggregates is very low and this is the main reason for the low cohesion of the whole coating. In order to improve the cohesion of the coatings more melted particles must be obtained. This means that the penetration of the droplets into the hot core of the plasma must be improved.

Crystalline Phases

The diffraction patterns obtained for the organic precursor dried at 100 °C (Fig. 6a), calcined at 500 °C (Fig. 6b) and plasma sprayed (Fig. 6c) are shown in Fig. 6. The pattern obtained at 100 °C shows diffraction peaks related to reaction products from the chemical precursors but no diffraction peaks of hydroxyapatite are detected. When those precursors were calcined at 500 °C only HA peaks are resolved.

The pattern of the sprayed coating depicted in Fig. 6c is a well resolved HA pattern. No peaks related to undecomposed precursors are observed. Diffraction peaks related to the decomposition of the HA, as tretracalcium phosphate and calcium oxide, are detected. The presence of this decomposition phase is related to the high temperatures reached by the substrate during spraying, as mentioned before. The presence of those decomposition phases must be avoided because they can have a negative effect on the biocompatibility of the coating and its stability in the body[11,12].

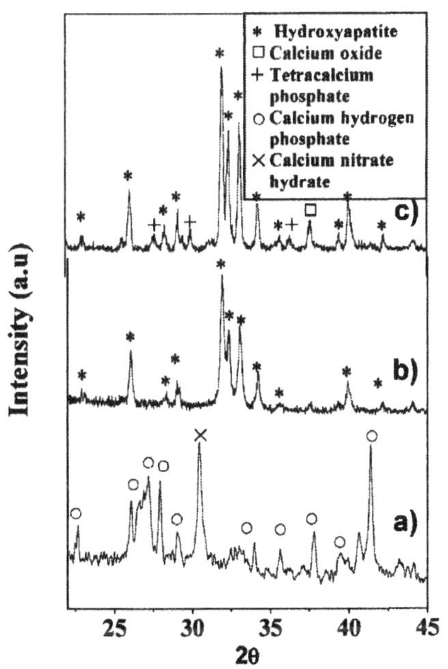

Figure 6: XRD patterns of a) Organic precursor dried at 100 °C, b) Organic precursor calcined 1 h at 500 °C and c) sprayed coating.

CONCLUSIONS

In this work solution precursor plasma spray (SPPS) technique has been used to deposit porous hydroxyapatite coatings on Ti6AL4V substrates using an organic precursor.

Microstructure analysis performed with TEM/SEM showed coatings formed by a combination of features characteristics of the SPPS technique, i.e. melted particles, hollow spheres and nanostructured aggregates. The presence of an interface reaction layer is also observed. This layer appears as the consequence of the high temperature reached during spraying.

The main crystalline phase detected in the coatings by X ray diffraction analysis is hydroxyapatite. Other minor peaks are also detected. These peaks are related with crystalline phases resulting from the decomposition of the hydroxyapatite.

For future work the temperature of the substrate must be controlled to avoid the formation of thick interface reaction layers and the decomposition of the Hydroxyapatite in order to obtain a high quality coating. In order to improve the cohesion of the coating more melted particles are desired in the coating. To obtain this goal the spraying parameters and the pressurized injection system must be optimized.

ACKNOWLEDGMENTS

The authors thank Dr. S. Hao, Dr. L. Pershin and Mr. T. Lee for their help with the different thermal spray techniques.

Eugenio Garcia wishes to acknowledge to National Secretary of Education and Universities of Spain and European Social Fund for financial sponsorship.

REFERENCES

[1]J. Furlong and J. F. Osborn, Fixation of Hip Prostheses by Hydroxyapatite Ceramic Coatings, *J. Bone Joint Surg*, **73B**, 741-745, (1991).

[2]G. T. Geeskink, Experimental and Clinical Experience with Hydroxyapatite Hip Implants, *Orthopedics*, **12**, 1239-1242, (1989).

[3]J. P. Collier, V. A. Surprenant, M. B. Mayor, M. Wrona, R. E. Jensen and H. P. Surprenant, Loss of Hydroxyapatite Coating on Retrieved Total Hip Components, *J. Arthroplasty*, **8**, 389-393, (1993).

[4]N. P. Padture, K. W. Schlichting, T. Bathia, A. Ozturk, B. Cetegen, E. H. Jordan, M. Gell, S. Jiang, T. D. Xiao, P. R. Strutt, E Garcia, P. Miranzo, M. I. Osendi, Towards Durable Thermal Barrier Coatings with Novel Microstructures Deposited by Solution Precursor Plasma Spray, *Acta Mater.*, **49**, 2251-2257, (2001).

[5]L. Xie, X. Ma, E. H. Jordan, N. P. Padture, D. T. Xiao and M. Gell, Deposition of Thermal Barrier Coatings using the Solution Precursor Plasma Spray Process. *J. Mater. Sci.*, **39**, 1639-1646 (2004).

[6]L. Gan and R. Pilliar, Calcium Phosphate Sol-Gel-derived Thin Films on Porous-Surfaced implants for enhanced osteoconductivity. Part 1: Synthesis and Characterization, *Biomaterials*, **25**, 5303-5312, (2004)

[7]D. M. Liu, T. Troczynski, Wenjea and J. Tseng. Water Based Sol-Gel Synthesis of Hydroxyapatite: Process Development, *Biomaterials*, **22**, 1721-1730 (2001).

[8]K.A. Gross, N. Ray, M. Rokkum. The Contribution of Coating Microstructure to Degradation and Particle Release in Hydroxyapatite Coated Prostheses. *J Biomed. Mater. Res.*, **63**, 106-114 (2002).

[9]Huaxia, C. B. Ponton and P. M. Marquis. Microstructural Characterization of Hydroxyapatite Coating on Titanium. *J. Mater. Sci. Mater. Med.*, **13**, 261-265, (1998).

[10]L. Gan, J. Wang and R. M. Pilliar. Evaluating Interface Strength of Calcium Phosphate Sol-Gel derived Thin Films to Ti6Al4V Substrate, *Biomaterials*, **26**, 189-196, (2005).

[11]S. R. Radin and P. Ducheyne. Plasma Spraying Induced Changes of calcium Phosphate Ceramic Characteristics and the Effect on in Vitro Stability. *J. Mater. Sci. Mater. Med.*, **3**, 33-42 (1992).

[12]R. Z. LeGeros. Biodegradation and Bioresorption of Calcium Phosphate Ceramics. *Clin. Mater.*, **14**, 65-68 (1993).

VISIBLE–LIGHT PHOTOCATALYTIC FIBERS FOR INACTIVATION OF PSEUDOMONAS AERUGINOSA

P. G. Wu[1], R. C. Xie[1], J. Imlay[2], and J. K. Shang[1*]
[1]Department of Materials Science and Engineering
[2]Department o f Microbiology
University of Illinois at Urbana Champaign, Urbana, IL 61801 (USA)
*Corresponding author, e-mail: jkshang@uiuc.edu

ABSTRACT

Antimicrobial ceramic fibers based on a visible-light photocatalyst were made by templating on activated carbon glass fibers. Crystalline photocatalyst nanoparticles were grown in the interconnected nanopore system of the activated carbon. Removal of the carbon template in the subsequent calcination resulted in nanoporous ceramic fibers. When the fibers were immersed in suspensions of *Pseudomonas aeruginosa* under visible light illumination, the survival ratio of the cell was found to decrease with the time of exposure to the visible light. The kinetics of the bacterial killing was of the first order, reaching more than 40% bacterial killing in 30 min.

1. INTRODUCTION

Photocatalytic inactivation of microorganism was first demonstrated for titanium dioxide in 1985 [1]. Since then, various organisms have been inactivated, such as bacteria [1-4, 5-6], bacterial and fungal spores [7], and algae [8]. The applications of photocatalytic killing have ranged from disinfection of air and water to the sterilization of hospital utensils [2-4]. For water treatment, a fine TiO_2 powder is often suspended in water and illuminated with an ultraviolet (UV) light source [4]. This kind of UV-TiO_2 system suffers from two major problems: the high cost and potential hazards of UV light source, and extra cost of powder recovery. To address the problem with the UV light source, much effort has been devoted to develop visible light photocatalysts [9, 10].

Despite numerous reports of visible light activity in various photocatalysts, very few studies have focused on photocatalytic killing of pathogens by visible light. Yu and co-workers [11] reported the visible-light-induced bactericidal effect of sulfur-doped nanocrystalline TiO_2. Under their experimental conditions, the killing ratio of Micrococcus lylae (gram-positive) was *ca.* 36% after 30 min light illumination.

Recently, we reported the synthesis of nitrogen doped titanium oxide (i. e., TiON) thin films by ion beam assisted vapor phase deposition process. A characteristic decreasing trend in bandgap values of the films was observed within a certain range of increasing dopant concentrations. As a result of bandgap narrowing, optical absorption of TiON was extended well into visible light region [12]. However, the thin film demonstrated rather low photocatalytic activity because of the low surface area.

To increase the contact efficiency and avoid extra cost of photocatalyst recovery, nanoporous fibers were made from nitrogen-doped titanium oxide photocatalyst. When illuminated by visible light, the fibers were found to kill *Pseudomonas aeruginosa.* To our knowledge, this is the first report of visible light-induced photocatalytic inactivation of *P. aeruginosa*. The killing rate followed the first order kinetics. After 30 min illumination, the survival ratio of *Pseudomonas aeruginosa* was reduced to less than 60%.

2. EXPERIMENTAL

Synthesis. Reagent grade titanium tetraisopropoxide (TTIP 98+%), tetramethylammonium hydroxide (TMA, 25% in methanol), and ethyl alcohol were purchased from Aldrich. These reagents are used as received without further purification. A mixture of TTIP and TMA (mol ratio 4:1) was first made as the precursor for nitrogen-doped titanium oxide (TiON). The mixture was then used to soak the activated carbon glass fiber template for 24 h. The hydrolysis of all precursors was initiated by exposure to the moisture in air and gelation was allowed for 24 h at room temperature. After wash and drying, fine crystallites of TiON were finally obtained by calcination at 400°C in air for 4 h, followed by removal of carbon at 550 °C in air at a constant heating rate.

Characterization. Fibers were observed with a JEOL JSM-7000F Field Emission Analytical Scanning Electron Microscope at 15 kV. The samples were characterized by transmission electron microscopy (TEM). A JEOL 2010 F electron microscope operating at 200 kV was used for the TEM investigations. Plan-view film specimens were prepared by removing the films from the substrate with a razor blade and suspending them in ethanol. This suspension was then dispersed on a holey carbon film supported by a copper grid. The analysis of surface area, pore size distribution and pore volumes was carried out with an Autosorb-1 volumetric sorption analyzer.

Bacterial culture. *Pseudomonas aeruginosa* ATCC 10145 were inoculated each time from an agar plate into a 4 ml liquid nutrient broth medium. The cells were grown aerobically in the medium placed on a rotary shaker at 37°C for 18 h. Cells were harvested from overnight culture by centrifugation for 5 min at 277 K and 6000 rpm, washed twice using a buffer solution (0.05 M KH_2PO_4 and 0.05 M K_2HPO_4, pH 7.0), then resuspended and diluted in buffer prior to the use for sterilizing experiments. The initial cell concentration was in the magnitude of 10^7 cfu/ml, determined by a viable count procedure on agar plates after serial dilutions. All solid or liquid materials have been autoclaved for 30 min at 121°C before use.

Photocatalytic inactivation and cell viability assay. The experimental set-up for a static bactericidal testing is shown schematically in Figure 1. At starting time, aliquot of 3-ml *P. aeruginosa* cell suspension was pipetted onto a sterile 60x15mm petri dish, with the fiber photocatalysts placed in the bottom. The amount of cell suspension was shallowly over the top of fibers (fiber thickness is 0.28mm). The photocatalyst applied in each test is about 1.0 gram TiON per liter of cell suspension. The covered petri dishes were illuminated by a metal halogen desk lamp equipped with a glass filter. Zero light intensity was detected below 400 nm. The light intensity striking the cell suspensions was *ca.* 1.6 mW/cm^2, as measured by a Multi-Sense MS-100 optical Radiometer (UVP, U.S.A). The temperature increased from room temperature to 40°C under illumination. The suspension was only stirred at the moment of sample withdrawal.

At regular time intervals, 20μL of aliquots of the irradiated cell suspensions were withdrawn in sequence. After appropriate dilutions in buffer solution, aliquots of 20μL together with 2.5 ml top agar was spread onto an agar medium plate and incubated at 37°C for 24 h. The number of viable cells in terms of colony-forming units was counted. Analyses were in duplicates and control runs were carried out each time under the same illumination conditions, but without any photocatalytic materials in the cell suspension.

3. RESULTS AND DISCUSSION

3.1 Templated Growth of Photocatalyst

The initial activated carbon glass fiber template contained interconnected pore system as shown in Fig. 2. Nitrogen adsorption analysis indicated that the average pore dimension in the activated carbon was 2 nm. Upon impregnation by the TiON precursor and crystallization, the fiber consisted of a composite structure in which carbon and oxide nanoparticles are intertwined. After removal of the carbon template, a nanoporous network of TiON photocatalyst was formed. The average pore size in the photocatalyst was 4 nm as measured from the nitrogen adsorption. From the high resolution transmission electron images, the average photocatalyst particle size was determined to be about 7 nm, greater than the initial pore dimension. The difference was believed to result from particle growth when the carbon template was gradually removed by the high temperature oxidation process.

3.2. Fiber Morphology

Since the template fibers were available as nonwoven fabrics, the photocatalyst fibers formed open networks as shown in Fig. 3. The fiber network created large pores of tens to hundreds of micrometers (Fig. 3A). These large pores are desirable both for improving the contact efficiency between the bacteria and TiON photocatalyst, and for reducing the pressure drop across the fabric in a dynamic reactor. At a higher magnification in Figure 3B, fine particles of TiON photocatalyst were clearly visible. X-ray diffraction analysis and electron diffraction confirmed that the particles were crystalline and assumed anatase structure.

3.3. Inactivation of *P. aeruginosa*

In this study, *P. aeruginosa* was chosen as the indicators for the photosterilization activity of visible-light photocatalysts for several reasons. First of all, *Pseudomonas aeruginosa*, is one of the most formidable biofilm-forming organisms. Biofilms are thought to be the predominant living structure for bacteria in natural environments, and a cause of human infections [13]. Secondly, this gram-negative, rod-shaped bacterial cell is ubiquitous in nature and is an opportunistic pathogen in humans. It survives on very little nutrient and is difficult to treat because it possesses several important antibiotic resistance mechanisms.

The effect of the photocatalyst on the viability of *P. aeruginosa* cells was studied by exposing the cells suspended in buffer solution to a visible light for varying time intervals. Data are presented as changes in the survival ratio, N_t/N_0, over the illumination time (N_0 and N_t are the number of colony-forming units at the initial and each following time interval, respectively. In Figure 4, sterilization tests indicated that TiO_2 immobilized on active carbon glass fiber virtually has no bactericidal effect under visible light illumination, which formed an overlapping straight line when compared to the control. On the contrary, the bactericidal function of TiON fibers started upon the first time interval, 10 min illumination, and became more and more evident with time. An overall trend of decrease in the survival ratio was observed for *P. aeruginosa* cells treated by TiON fibers, with a survival fraction of 57% at 30 min illumination, and 32% at 60 min. On a semilog plot (Fig. 4), the killing curve appeared linear, or was of the first order. When such an inactivation rate is compared to TiO_2 irradiated under visible light, nitrogen doping has evidently enabled photo-sterilization without the requirement of UV light.

For the studies of photocatalytic inactivation, another gram-negative cell *E .coli* has been extensively studied [1-4, 5-6, 11], because it is easily grown in a nutrient broth at lab and the wild type *E .coli* is not a pathogen to human. Only a few reports have inactivated *P. aeruginosa* cell using TiO_2 under UV irradiation. Seven and coworkers reported the inactivation of *P. aeruginosa* irradiated with a 400 W sodium lamp to simulate solar radiation for various time periods [15]. An efficient microbicidal effect of TiO_2 was detected and the strain of *P. aeruginosa* bacteria was destroyed in 40 min. In a study conducted by Robertson et al. when *P. aeruginosa* was exposed to TiO_2 and UVA light, a substantial decrease in bacterial numbers was observed and the survival is about 0.0001 after irradiation 2h [16]. Compared to these studies, our study used visible light rather than UV to activate the photocatalytic process in killing *P. aeruginosa.* Although the experimental conditions differed greatly, the killing rates of the photocatalytic processes are comparable.

The primary working species for photocatalytic killing is widely believed to be hydroxyl radicals formed at the surface of photocatalyst [11, 15]. Amezaga-Madrid et al. observed in TEM the ultrastructural alteration on *P. aeruginosa* cell by photocatalytic TiO_2 thin films. The alteration was mainly related to abnormal cell division and the cell membrane [17]. In the present reaction system, the opening of the fabric structure is much larger than a typical *P. aeruginosa* cell which is no more than 1-2 micrometers. Therefore, TiON particles, and the oxidative species (mainly hydroxyl radicals) that are produced by the irradiation of the photocatalyst, should have ample opportunity to come in direct contact with the microorganisms. It is well known that reactive oxygen species can readily attack the microbial external and internal structures. In a parallel study, we have detected oxidative species produced by the TiON fibers and the observed severe oxidative damages on *E. coli* cell wall, cell membrane and DNA molecules by visible-light photocatalysis [18]. An example of the damaged E. coli cell is shown in Figure 5. There can be seen large numbers of rumples and high degree of disordered configurations on the cell wall and the cell membrane. Holes and/or deep collapse appeared in nearly every cell. In contrast, the surfaces of untreated *E. coli* cells are continuous and damage-free [18]. We believe the mechanisms of killing P. aeruginosa by visible-light induced photocatalysis with TiON fibers are similar to that of killing E. coli with the same materials and conditions. Namely, the photocatalytic damages on the two gram-negative bacteria cells take place both externally and internally, as morphological changes on the bacteria cell wall and the cell membrane, together with DNA alterations [18].

4. CONCLUSION

Porous fibers of TiON photocatalyst were made by templated growth of TiON particles in the pore system of activated carbon glass fibers. The fibers demonstrated photocatalytic inactivation of P. *aeruginosa* under visible light irradiation. Since a fiber form of the photocatalyst saves the extra cost associated with the recovery of catalyst particles and TiON can take advantage of the visible light from the solar energy, these visible-light photocatalyst fibers present great promise in providing alternative antimicrobial solutions to the traditional chlorination for water disinfection.

ACKNOWLEDGMENTS

This material is based upon work supported by the Center of Advanced Materials for the Purification of Water with Systems, National Science Foundation, under Agreement Number CTS-0120978. The authors thank Dr. Z. Yue and Prof. J. Economy for providing the activated carbon glass fiber templates. XRD, TEM and SEM were performed at CMM center of the Frederick Seitz Materials Research Laboratory (partially supported by the U.S. Department of Energy under grant DEFG02-91-ER45439).

REFERENCES

1. T. Matsunaga, R. Tomoda, T. Nakajima and H. Wake, "Photoelectrochemical sterilization of microbial cells by semiconductor powders", *FEMS Microbiol Lett* **29**, 211-214 (1985).
2. Y.-S. Choi, B.-W. Kim, "Photocatalytic disinfection of E coli in a UV/TiO_2-immobilised optical-fibre reactor", *J. of Chemical Technology & Biotechnology* **75**, 1145-1150 (2000).
3. Y. Ohko, Y. Utsumi, T. Tatsuma, C. Niwa, Y. Kubota, K. Kobayakawa, Y. Satoh, and A. Fujishima, "Self-sterilizing and self-cleaning of silicone catheters coated with TiO_2 photocatalyst thin films", *J Biomed Mater Res.* **58**, 97-101 (2001).
4. N. Huang, Z. Xiao, D. Huang, C. Yuan, "Photochemical disinfection of Escherichia coli with a TiO_2 colloid solution and a self-assembled TiO_2 thin film" *Supramolecular Sci.* **5**, 559-564 (1998).
5. Y. Koizumi, J. Nishi, M. Taya, "Photosterilization of Escherichia coli Cells Using Iron-Doped Titanium Dioxide Particles" *J. Chem. Eng. Jap.* 35 (2002) 299-303.
6. F. M. Salih, "Enhancement of solar inactivation of Escherichia coli by titanium dioxide photocatalytic oxidation", *J. of Applied Microbiology* 2002, 92, 920-926.
7. R.W. Armon and P. G. Bettane "Disinfection of Bacillus spp. Spores in drinking water by TiO_2 photocatalysis as a model for Bacillus anthracis"*Water Sci. Technol.: Water Supply* **4**, 7-14 (2004).
8. C. Linkous, G. Carter, D. Locuson, A. Ouellette, D. Slattery, and L. Smitha, "Photocatalytic inhibition of algae growth using TiO_2, WO_3, and cocatalyst modifications" *Environ. Sci. Technol* **36**, 3412-3419 (2002).
9. Y. Sakatani, J. Nunoshige, H. Ando, K. Okusako, H. Koike, T. Takata, J. N. Kondo, M. Hara, and K. Domen "Photocatalytic decomposition of acetaldehyde under visible light irradiation over La^{3+} and N co-doped TiO_2" *Chem. Lett.* **32**, 1156-1157 (2003).
10. Y. Sakatani, J. Nunoshige, H. Ando, K. Okusako, H. Koike, T. Takata, J. N. Kondo, M. Hara, and K. Domen "Metal ion and N co-doped TiO_2 as a visible-light photocatalyst" *J. Mater. Res.* **19**, 2100-2108 (2004).
11. J. C. Yu, W. Ho, J. Yu, H. Yip, P. Wong and J. Zhao "Efficient visible-light-induced photocatalytic disinfection on sulfur-doped nanocrystalline titania", *Environ. Sci. Technol* **39**, 1175-1179 (2005).
12. P. G. Wu, C. H. Ma, J. K. Shang *J. Appl. Phys. A* **81**, 1411-1417 (2005).
13. B. R. Boles, M. Thoendel, and P. K. Singh "Self-generated diversity products 'insurance effects' in biofilm communities", *Proceedings of the National Academy of Sciences of the USA*, 101, 16630-16635 (2004).
14. W. Harm "Biological effects of ultraviolet radiation", Cambridge University Press, New York, p.45 (1980).

15. O. Sevena, B. Dindara, S. Aydemirb, D. Metinb, M. A. Ozinelb and S. Icli "Solar photocatalytic disinfection of a group of bacteria and fungi aqueous suspensions with TiO_2, ZnO and Sahara desert dust", *J. Photochem. Photobio. A*, **165**, 103-107 (2004).
16. Jeanette M.C. Robertsona, Peter K. J. Robertsona and Linda A. Lawton "A comparison of the effectiveness of TiO_2 photocatalysis and UVA photolysis for the destruction of three pathogenic micro-organisms", *J. Photochem. Photobio. A*, **175**, 51-56 (2005).
17. P. Amezaga-Madrid, R. Silveyra-Morales, L. Cordoba-Fierro, G. V. Nevarez-Moorillon, M. Miki-Yoshida, E. Orrantia-Borunda and F. J. Solis "TEM evidence of ultrastructural alteration on Pseudomonas aeruginosa by photocatalytic TiO_2 thin films", *Photochem. Photobiol. B* **2003**, 70, 45-50.
18. P. Wu, R. Xie, J. Imlay, J. K. Shang "Photocatalytic Inactivation of *Escherichia coli* by a Modified TiO_2 under Visible-light Illumination and the Killing Mechanisms", Unpublished work, University of Illinois, 2005.

FIGURES AND CAPTIONS

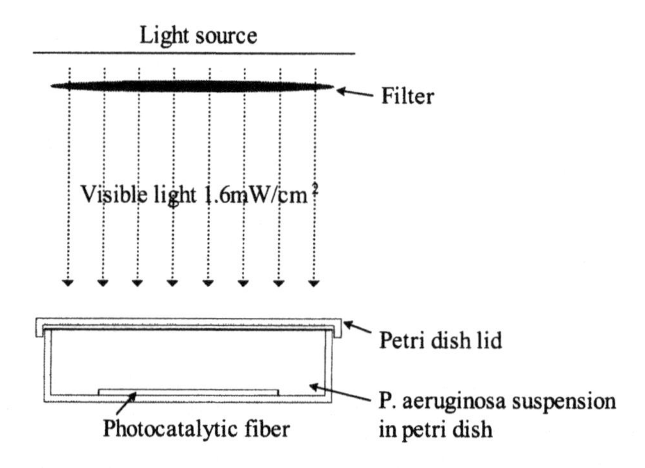

Figure 1 Schematic illustration of the photocatalytic testing setup.

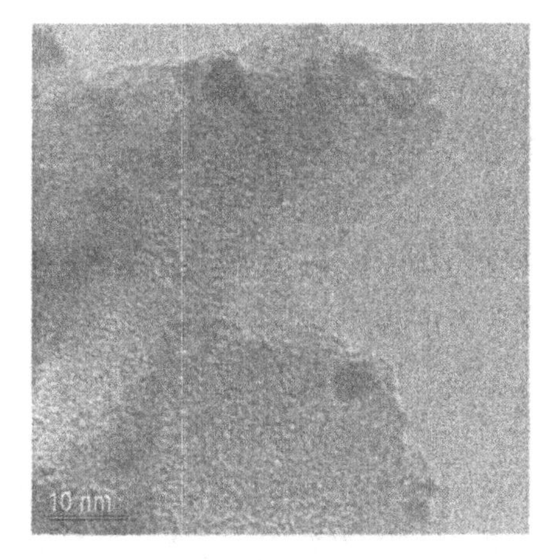

Figure 2 High resolution transmission electron microscopy image of the pore system in activated carbon glass fiber template.

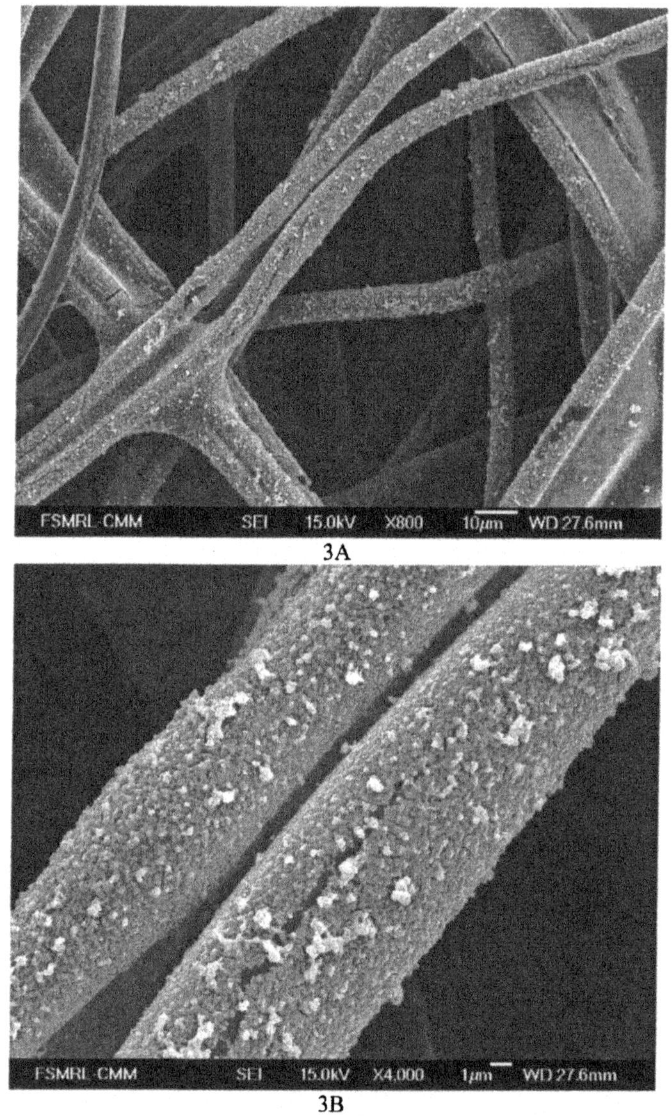

Figure 3 Scanning electron micrographs of TiON fiber, (A) showing its open fabric structure; (B) showing the incorporation of TiON particles.

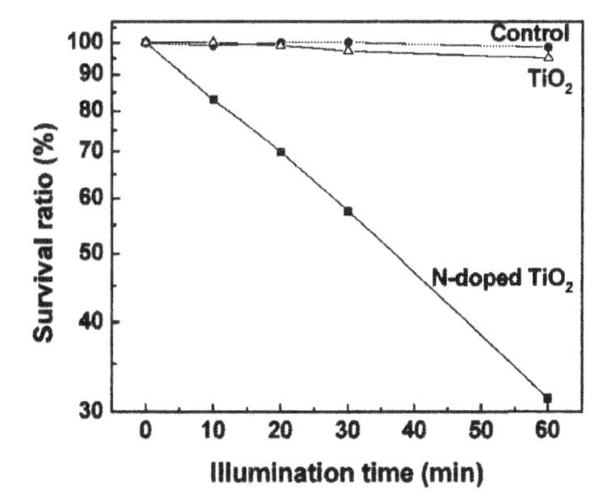

Figure 4 Effects of pure TiO_2 and TiON fibers on *P. aeruginosa* survival ratio. Initial concentration 10^7 cfu/ml. Inactivation of *P. aeruginosa* by tungsten halogen light with and without photocatalyst addition.

Figure 5 SEM image showing the oxidative damages on *E. coli* cells upon visible light illumination in the presence of a TiON-based fiber after 120 min. The cell suspension had an initial concentration *ca.* 10^9 cfu/ml. Untreated *E. coli* cells have a damage-free surface morphology, see Ref. [18].

PRECIPITATION MECHANISMS OF HYDROXYAPATITE POWDER IN THE DIFFERENT AQUEOUS SOLUTIONS

Yu-Chen Chang, Tzer-Shin Sheu
Department of Materials Science and Engineering
I-Shou University, Kaohsiung, Taiwan

ABSTRACT

Hydroxyapatite powder was prepared in the different aqueous solutions, by varying the concentrations of PO_4^{3-}, Ca^{2+} and OH^-. Sugar, KOH and glycerol were added into the aqueous solutions to observe their effects in the formation of hydroxyapatite precipitate at different pHs in this study. With a long aging treatment up to 24h at 25-60℃, a plate-like hydroxyapatite precipitate appeared in the aqueous solutions with KOH, glycerol additives, and without any additives. However, a granular hydroxyapatite precipitate was obtained in the aqueous system containing a sugar additive. From microstructure observations, plate-like hydroxyapatite precipitates had an aspect ratio up to 5 and a width of 60nm. From X-ray diffraction method, the size of submicrostructure in these aged precipitates was 20Å approximately, which was like independent of solution additives and aging time.

INTRODUCTION

Hydroxyapatite (HA,$Ca_{10}(PO_4)_6(OH)_2$) has been known to have good biological compatibility and osteoconduction for orthopedic applications.[1] Typical synthesizing methods for obtaining hydroxyapatite(HA) powders are solid-state reaction, chemical co-precipitation , hydrothermal,[2-3] sol-gel method,[4] and coating methods.[5] When a chemical co-precipitation was selected to obtain HA powders, Ca^{2+} and PO_4^{3-} were firstly dissolved in the water, and then the processing factors, such as pH value, supersaturation and reaction temperature, were appropriately controlled to obtain different types of calcium phosphate precipitates. Among these processing factors, pH value, solution additive, and aging time are believed to play a very important role in the formation of hydroxyapatite precipitates. Therefore, solution additives like sugar, KOH, and glycerol are chosen to observe whether these additives will significantly affect the morphology of HA precipitate at different temperatures, pH value, and aging time in this study.

EXPERIMENTAL PROCEDURES

(A) Synthesis of hydroxyapatite powders

Chemicals $Ca(OH)_2$ and H_3PO_4 were added into de-ionized water to form a 0.5M of solid particle dispersed solution and a 0.3M of clear aqueous solution, separately.[6-7] Without any other solution additive, a clear 0.3M of H_3PO_4 solution was droplike added into a continuously stirred 0.5M of $Ca(OH)_2/H_2O$ solution at 25-60℃. After these two solutions were well reacted, a batch of solution was collected to observe microstructures and phase existence of its solid contents in each specific time period.

With a solution additive, a various concentration of KOH, sugar, or glycerol solution, and a clear 0.3M of H_3PO_4 solution were simultaneously droplike added into a continuously stirred 0.5M of $Ca(OH)_2/H_2O$ solution at 25-60℃, to observe the effect of each solution additive in the formation of hydroxyapatite. After well reacted, solid precipitates were collected in each time period, to observe their microstructures and phase existence.

(B) Phase existence and microstructure observations

Morphologies of solid contents or precipitates were observed either by Field Emission Gun Scanning Electron Microscopy (FE-SEM), or by a transmission electron microscope (FEI Tecnai $G^2$20 S-TWIN), which was operated at 200 kV. An TEM specimen was prepared by using a carbon-coated copper grid to collect solid contents or precipitates from each well reacted solution.

X-ray diffraction method was used to determine the phase existence of solid contents or precipitates, in which the X-ray diffraction was operated at a voltage of 35 KV and a beam current of 30 mA. Except for phase existence and microstructure, Fourier transform infrared spectroscopy (FT-IR) was also used to identify the presence of some chemical bonds.

RESULTS AND DISCUSSION

(A) Without any solution additive

Without any solution additive, X-ray diffraction patterns of solid contents or precipitates in the well reacted $Ca(OH)_2$/ H_3PO_4 solution as a function of aging time is shown in Fig. 1. From Fig. 1(a), it indicates that hydroxyapatite precipitate evolves before an aging time of 10 min, and no any trace of $Ca(OH)_2$ diffraction peaks is observed. As the aging time increases, there are no significant differences of X-ray diffraction patterns among these different aged specimens, which can be seen from Figs. 1(a)-(f).

Table I. Processing variables in the 0.5M $Ca(OH)_2$/0.3M H_3PO_4 chemical coprecipitation system

Sample index	additive	Temperature (℃)	Aging time (h)
T25-0-0.2	no	25	0.2
T25-0-1	no	25	1
T25-0-3	no	25	3
T25-0-5	no	25	5
T25-0-10	no	25	10
T25-0-24	no	25	24
T25-S-5	0.73M sugar	25	5
T25-S-10	0.73M sugar	25	10
T25-S-24	0.73M sugar	25	24
T25-K-5	0.1M KOH	25	5
T25-K-10	0.1M KOH	25	10
T25-K-24	0.1M KOH	25	24
T25-G-5	0.1M glycerol	25	5
T25-G-10	0.1M glycerol	25	10
T25-G-24	0.1M glycerol	25	24
T60-0-5	no	60	5
T60-0-10	no	60	10
T60-0-24	no	60	24
T60-S-5	0.73M sugar	60	5
T60-S-10	0.73M sugar	60	10
T60-S-24	0.73M sugar	60	24
T60-K-5	0.1M KOH	60	5
T60-K-10	0.1M KOH	60	10
T60-K-24	0.1M KOH	60	24
T60-G-5	0.1M glycerol	60	5
T60-G-10	0.1M glycerol	60	10
T60-G-24	0.1M glycerol	60	24

The corresponding SEM micrographs of these different aged precipitates are shown in Fig. 2. With an aging time of 5-10h, the aspect ratio of HA precipitates is 2-4, as shown in Figs. 2(a)& (b). As the aging time increases, the aspect ratio of HA precipitates increases to 5 or more, which can be easily seen from the sample T25-0-24 shown in Fig. 3(c).

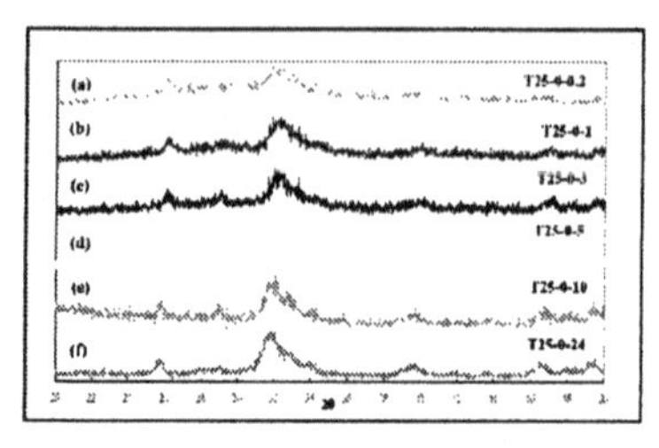

Fig. 1 XRD patterns of solid contents or precipitates in the 0.5M $Ca(OH)_2$/0.3M H_3PO_4 system without any solution additives after being aged at room temperature for (a) 0.2h, (b) 1h, (c) 3h, (d) 5h, (e) 10h, and (f) 24h.

One of aged specimens without any solution additive shows its transmission electron micrograph (TEM) in Fig. 3. This TEM micrograph again reveals HA precipitates to be plate-like or acicular, which is very similar to the SEM micrograph shown in Fig. 2(a).

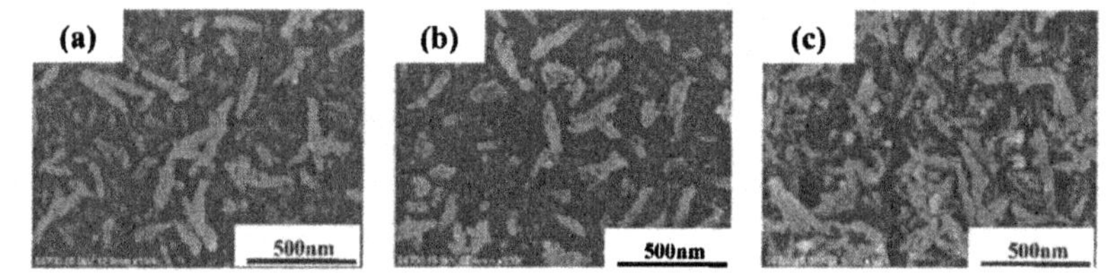

Fig.2 SEM micrographs for HA precipitates in the 0.5M $Ca(OH)_2$/0.3M H_3PO_4 system without any solution additives after the aging time of (a) 5h, (b) 10h, and (c) 24h.

Fig. 3 Transmission electron micrograph for sample T25-0-5.

Some FTIR spectra for the aged specimens with the solution additives are shown in Fig. 4. It shows the absorption peaks of OH^- bond appear at the wavenumber (λ^{-1}) of 793 cm^{-1}, 3363 cm^{-1} and 3571 cm^{-1}, PO_4^{3-} bond at 960 cm^{-1}, and 1028 cm^{-1}. These characteristic absorption peaks further prove those solid contents or precipitates to be hydroxyapatite (HA), but not other phase existed.

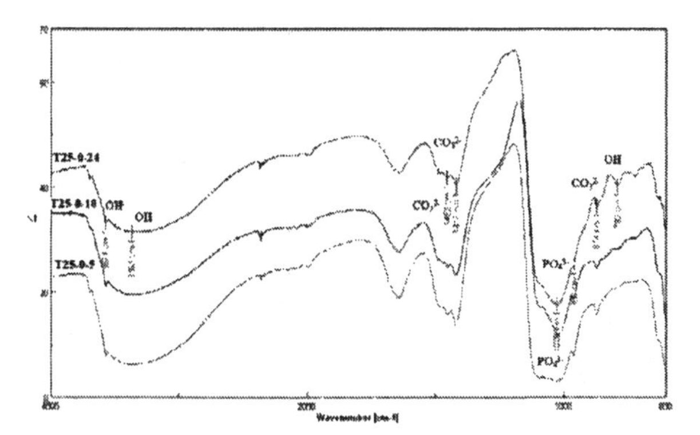

Fig. 4 FTIR spectra for the aged specimens without any solution additive. Sample indices are referred to Table I.

(B) Synthesis of hydroxyapatite powders with a solution additive at 25℃

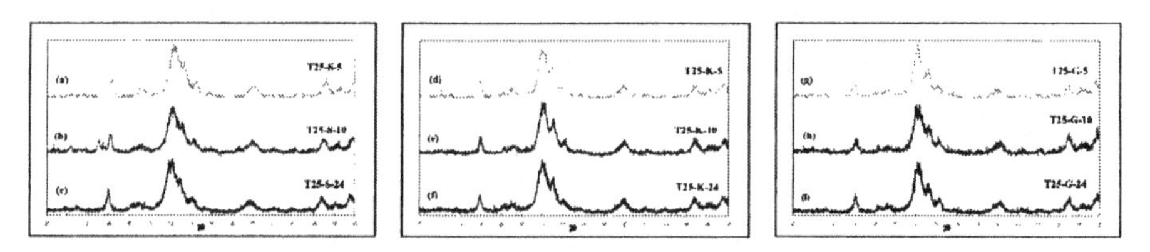

Fig. 5 X-ray diffraction patterns for the aged specimens with the solution additives: (a)-(c) sugar, (d)-(f) KOH, and (g)-(i) glycerol. Sample indices are marked on each X-ray diffraction pattern, in which sample indices are referred to Table I.

X-ray diffraction patterns for the aged specimens with different solution additives are shown in Fig. 5. With a different aging time and a different solution additive, there are no significant differences among these aged specimens, all with the same HA diffraction patterns. The corresponding SEM micrographs are shown in Fig. 6. Except for the specimens with a sugar additive, HA precipitates in the other aged specimens are all with the elongated plate-like structure, as shown in Figs. 6(d)-(i). HA precipitates in the aged specimens with a sugar additive are granular, as shown in Figs. 6(a)-(c).

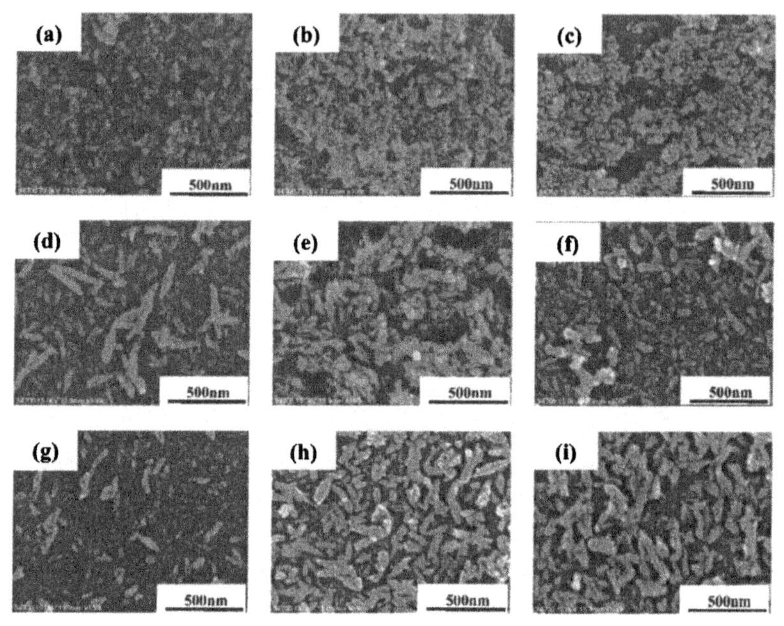

Fig. 6 SEM micrographs for the aged specimens (a) T25-S-5, (b) T25-S-10, (c) T25-S-24, (d) T25-K-5, (e) T25-K-10, (f) T25-K-24, (g) T25-G-5, (h) T25-G-10, and (i) T25-G-24. Sample indices are referred to Table I.

(C) Synthesis of hydroxyapatite powders at 60℃

X-ray diffraction patterns for the aged specimens with or without any solution additives are selectivity shown in Fig. 7. All aged specimens contain a single HA phase. However, the aged specimens with a sugar additive have much broader X-ray diffraction peaks, as shown in Figs. 7(d)-(f). The corresponding SEM micrographs are shown in Fig. 9. The aged specimens with a sugar additive have much smaller HA precipitates, as shown in Figs. 9(d)-(f). The above results indicate that fine HA crystalline structures in the aged specimens with a sugar additive are consistent with their broader x-ray diffraction peaks shown in Figs. 7(d)-(f).

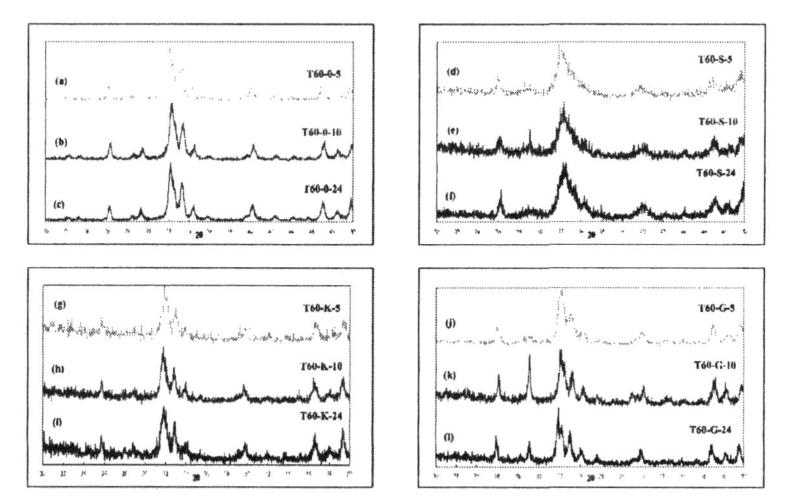

Fig. 7 X-ray diffraction patterns of different aged specimens (a) T60-0-5, (b) T60-0-10, (c) T60-0-24, (d) T60-S-5, (e) T60-S-10, (f) T60-S-24, (g) T60-K-5, (h) T60-K-10, (i) T60-K-24, (j) T60-G-5, (k) T60-G-10, and (l) T60-G-24 after being aged at 60℃.

By using Sherrer equation[8-10] and the half width of (002) diffraction peak for HA, the particle (or submicrostructure) size of each aged specimen is listed in Table II. Form the data listed in Table II, it indicates that the submicrostructure size is much smaller than the particle size obtained from SEM or TEM micrograph observation. Therefore, it is believed that each HA particle contains several submicrostructures. From TEM microstructure observation, the submicrostructures are observed in each HA particle (or plate-like structure). The submicrostructures size in TEM micrographs is ~20nm approximately, as shown in Fig. 8.

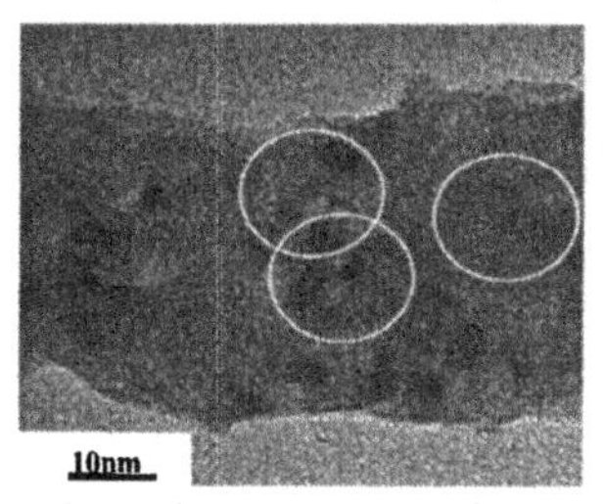

Fig. 8 TEM micrographs for sample T25-0-5

Table II. Size of submicrostructure in Different aged specimens calculated by Sherrer equation.*

Sample index	Size of submicrostructure (nm)
T25-0-5	17.0
T25-0-10	17.1
T25-0-24	25.4
T25-S-5	16.9
T25-S-10	17.0
T25-S-24	8.5
T25-K-5	17.0
T25-K-10	12.9
T25-K-24	8.5
T25-G-5	10.2
T25-G-10	20.2
T25-G-24	25.6
T60-0-5	17.0
T60-0-10	17.0
T60-0-24	17.0
T60-S-5	12.7
T60-S-10	17.0
T60-S-24	12.8
T60-K-5	17.1
T60-K-10	12.8
T60-K-24	17.0
T60-G-5	25.6
T60-G-10	17.1
T60-G-24	17.0

*From Sherrer equation, $D=0.9\lambda/(B\cos\theta)$, where D is the diameter of crystalline particle (or submicrostructure) ; λ, wavelength of radiation (X-ray); θ, angle of (002) diffraction peak; B, the radian of half-width of (002) diffraction.

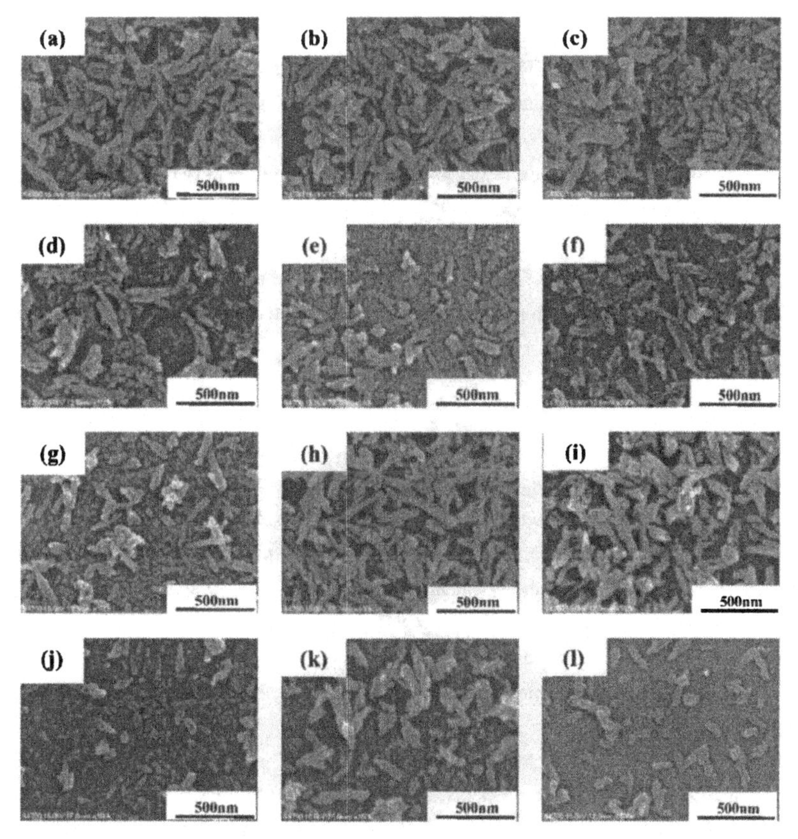

Fig. 9 SEM micrographs for aged specimens (a) T60-0-5, (b) T60-0-10, (c) T60-0-24, (d) T60-S-5, (e) T60-S-10, (f) T60-S-24, (g) T60-K-5, (h) T60-K-10, (i) T60-K-24, (j) T60-G-5, (k) T60-G-10, and (l) T60-G-24 after being aged at 60℃ for different time periods.

CONCLUSIONS

By using $Ca(OH)_2$ and H_3PO_4 as the initial chemicals, a single phase of hyproxyapatite (HA) precipitates or powders were prepared at 25-60℃ in an aqueous solution with or without any solution additives, such as sugar, KOH, and glycerol.

Without any solution additives, a plate-like or acicular HA precipitate existed in the aqueous solution at 25-60℃. However, HA precipitates obtained at 25℃ had much broader X-ray diffraction peaks, which implied a much smaller submicrostructure or grain size existed in the low-temperature processed powders.

With a solution additive like KOH or glycerol, HA precipitates showed a plate-like or acicular structure. However, HA precipitates in the aged specimens with a sugar additive were granular, but not plate-like.

From X-ray diffraction patterns, SEM and TEM micrograph observations, each granular or plate-like HA particle contained several submicrostructures with a size of 20nm approximately.

Except for microstructure observations and phase existence determination. FTIR spectra further indified O-H, P-O, and P-O-P bonds existing in all aged specimens. The presence of these bonds further confirmed these chemically coprecipitated powders were hyproxyapatite powders.

REFERENCES

[1]Hench LL. Bioceramics, from concept to clinics. *J. Am. Ceram. Soc.*, **74**, 1487 (1991).

[2]Changsheng Liu, Yue Huang, Wei Shen, and Jinghua Cui, "Kinetics of hydroxyapatite precipitation at pH 10 to 11," *Biomaterials*, **22**, pp.301-306, April, (2000).

[3]Shwu-Jen Jang, "A Study on Preparation of Hydroxyapatite Powders by Hydrothermal Method," *Master Thesis, Department of Chemical Engineering, Chung Yuan Christian University,* Taiwan, (1993).

4Wei-Lun Shieh, "Preparation and Related Property Evaluation of Hydroxyapatite Thick Film," *Master Thesis, Department of Materials Science & Engineering, I-Shou University,* Taiwan, (2002).

[5]Bressiani,A. H. De Campos, Magali "Comparative study of the synthesis and processing of Hydroxyapatite," *Key Engineering Materials*, 218-220, pp.171-174 (2002).

[6] K. Ioku, M. Yoshimura and S. Somiya, "Characterization of Fine Hydroxyapatite Powders Synthesized under Hydrothermal Conditions," *Sintering'87*, **2**, pp.1308-1313. Elsevier Applied Science, London, New York, Tokyo (1988).

[7] K. Ioku, M. Yoshimura, "Stoichiometric Apatite Fine Single Crystals by Hydrothermal Synthesis," *Phosp. Res. Bul.* ,**1**, pp.15-20 (1991).

[8]A. Osaka, Y. Miura, K. Takeuchi, "Calcium Apatite Prepared from Calcium Hydroxide and Orthophosphoric Acid," *Journal of Materials Science:Materials in Medicine*, **2**, pp.51-55 (1991).

[9]K. Ioku, S. Somiya, M. Yoshimura, " Hydroxyapatite Ceramics with Tetragonal Zirconia Particles Dispersion Prepared by HIP Post-Sintering" *Journal of Japan Ceramic Society*, **3**, pp.196-203 (1991).

[10]B. D. Cullity, "Elements of X-ray Diffraction," *Addison-Wesley Publishing Company, Inc.* pp.284-285 (1987).

CONVERSION OF BIOACTIVE SILICATE (45S5), BORATE, AND BOROSILICATE GLASSES TO HYDROXYAPATITE IN DILUTE PHOSPHATE SOLUTION

Wenhai Huang, Mohamed N. Rahaman, and Delbert E. Day, University of Missouri-Rolla, Department of Materials Science and Engineering, and Materials Research Center, Rolla, MO 65409

ABSTRACT

Compositional modification of bioactive 45S5 glass was performed by replacing the SiO_2 content partially or fully with B_2O_3 to form borosilicate and borate glasses. Its effect on the conversion of the glass to hydroxyapatite (HA) in dilute phosphate solution was investigated using kinetic, chemical, and structural techniques. A higher B_2O_3 content of the glass produced an increase in the conversion rate and a decrease in the pH of the solution. Particles of the borate glass (150–300 μm) were fully converted within 4 days, yielding pseudomorphic HA particles with a nanoscale structure. Silicate and borosilicate glass particles were only partially converted even after 70 days, forming a composite structure consisting of a SiO_2-rich core surrounded by a HA layer. Regardless of the composition, all the Na and B present in the glass particles dissolved into solution, but the Ca either reacted to from HA or remained in the unconverted SiO_2-rich core. The results are applicable to the development of bioactive glasses with controllable conversion rates to HA, which may provide a novel class of scaffold materials for bone tissue engineering.

INTRODUCTION

Since the report of its bone-bonding properties in 1971 by Hench *et al.* [1], the silicate-based bioactive glass codenamed 45S5, and referred to as Bioglass®, with a typical composition of 45% SiO_2, 24.5% Na_2O, 24.5% CaO, and 6% P_2O_5 (by weight), has been of primary interest for biological applications [2-4]. Later studies revealed that several silicate-based glasses and glass-ceramics also have the ability to enhance bone formation and bond to surrounding tissue [2-11]. An important characteristic of bioactive glasses is the time-dependent modification of the surface, leading to the formation of a HA layer that bonds to the surrounding tissue [12,13]. It has been suggested that the formation of an HA layer *in vitro* is indicative of a material's bioactive potential *in vivo* [14]. Key steps in the conversion of 45S5 glass to HA has been described by Hench [3,4], but many details of the chemical and structural changes that accompany the conversion are not clear. It is well established that that an initial step in the reaction is the formation of a SiO_2-rich gel on the 45S5 glass surface by ion exchange reactions. Further dissolution of ions from the glass and their diffusion through the SiO_2-rich gel layer, followed by the reaction between Ca^{2+} ions from the glass and PO_4^{3-} ions from the surrounding liquid, leads to the growth of HA on the gel layer.

More recently, the potential of borate glasses in biomedical applications has been explored [15-17]. One important rationale for the interest in borate-based bioactive glass is the lower chemical durability of some borate glasses, when compared to 45S5 glass, leading to an enhanced conversion rate to HA. A borate glass with the same composition as 45S5 glass but with all the SiO_2 replaced with B_2O_3, was investigated by Richard [18], who found that a HA layer not only formed on the borate glass surface upon immersion in a K_2HPO_4 solution at 37°C, but it also did so more rapidly than on 45S5 glass. The conversion of the borate glass to HA appeared to follow a process similar to that for 45S5 glass, but without the formation of a SiO_2-rich layer [19]. However, details of the reaction kinetics and mechanism, as well as the chemical and structural changes occurring during the conversion are not clear. In a preliminary *in vivo* experiment, particles of the borate glass (partially reacted in K_2HPO_4 solution to produce a HA surface layer) were found to promote bone formation more rapidly than 45S5 glass particles upon implantation into tibial defects in rats [18]. A borate glass with a composition somewhat similar to that used by Richard [18] was recently found to support the growth and differentiation of human mesenchymal stem cells [20]. Bioactive borate glasses have also been found to sinter more readily than 45S5 glass, enabling the formation of porous constructs for applications as scaffolds in tissue engineering [21].

The rapid conversion of some borate glasses to HA at near body temperature, the apparently favorable *in vitro* and *in vivo* reaction to cells and tissues, and the ability to form porous constructs appear promising, but a more detailed understanding the conversion and the cellular interaction is required to assess the potential of borate glasses for biomedical applications. In this work, the kinetics and mechanisms of the conversion to HA were addressed. The objective of the work was to investigate how compositional modification of bioactive 45S5 glass, by replacing the SiO_2 content partially or fully with B_2O_3 to form borosilicate and borate glasses, influenced the conversion of the glass to HA in a dilute phosphate solution at 37°C. Weight loss and pH changes, as well as chemical and structural changes accompanying the conversion, were studied.

EXPERIMENTAL

Glass Preparation

Four different glass compositions were used in the experiments (Table 1). One was the silicate-based 45S5 bioactive glass (designated in this work as 0B). Another glass (designated 3B), a so-called borate equivalent of 45S5 glass, had the same composition as 45S5 glass but all the SiO_2 was replaced with B_2O_3. The two remaining glasses had intermediate borosilicate compositions between 0B and 3B, with 1/3 and 2/3 of the SiO_2 molar concentration in 45S5 replaced with B_2O_3 (designated 1B and 2B, respectively).

Table 1. Compositions of the four glasses used in the experiments

Glass	Mole percent					Weight percent				
	Na_2O	CaO	B_2O_3	SiO_2	P_2O_5	Na_2O	CaO	B_2O_3	SiO_2	P_2O_5
0B	24.4	26.9	0	46.1	2.6	24.5	24.5	0	45.0	6.0
1B	24.4	26.9	15.4	30.7	2.6	24.0	23.9	17.0	29.3	5.8
2B	24.4	26.9	30.7	15.4	2.6	23.4	23.4	33.1	14.4	5.7
3B	24.4	26.9	46.1	0	2.6	22.9	22.9	48.6	0	5.6

The preparation of the glasses are described in detail elsewhere [22]. Briefly, the required quantities of SiO_2, H_3BO_3, $CaCO_3$, Na_2CO_3, and $NaH_2PO_4.2H_2O$ were heated in a platinum/rhodium crucible in air for 2 h at 1100°C for compositions 2B and 3B, at 1200°C for 1B, and at 1300°C for 0B. After casting, each glass was crushed in a hardened steel mortar and pestle to form particles, which were sieved through stainless steel sieves. Particles with sizes of 150–300 μm were used in most of the experiments performed to measure the reaction kinetics and to characterize the compositional and microstructural changes accompanying the conversion to HA. To better characterize microstructural and compositional changes across the particle cross-section, spheres of 3B glass, 800–1000 μm in diameter, and fibers of 0B glass, 200–300 μm in diameter, were also prepared.

Phosphate solution for conversion reaction

Conversion of the glasses to HA was accomplished *in vitro* by immersing them in a dilute phosphate solution. The solution was prepared by dissolving K_2HPO_4 (Reagent grade; Fisher Scientific, St. Louis MO) in deionized water (pH = 5.5 ± 0.1) to give a 0.02 M concentration of K_2HPO_4. The starting pH of the solution was 7.0 ± 0.1, approximately equal to the pH of human body fluid. The K_2HPO_4 concentration was chosen on the basis of the quantities of glass particles and solution used in the experiments (1 g glass in 100 cm^3 solution), with the requirement that sufficient PO_4^{3-} ions would be available in the solution to react with all the Ca^{2+} ions from the glass to form stoichiometric HA [$Ca_{10}(PO_4)_6(OH)_2$]. In fact, the K_2HPO_4 concentration was chosen to leave a small concentration (200–300 ppm) of residual PO_4^{3-} ions in the solution if the conversion reaction went to completion. This ensured that the Ca^{2+} content of the glass was the only compositional factor that could limit the completion of the reaction.

Weight loss and pH measurements

Conversion of the glasses to HA in the phosphate solution was accompanied by a weight loss, which was measured and used to monitor the kinetics of the conversion reaction. A reaction system consisting of 1.00 g glass particles (150–300 μm) in 100 cm^3 solution at 37 ± 2°C was used in all the conversion experiments. Prior to immersion, the glass particles were washed ultrasonically three times with deionized water, then twice with ethanol, and dried overnight at 90°C. During the conversion process, the system was removed from the oven at regular intervals and vibrated for 30 s each time to prevent the glass particles from sticking together. Weight loss measurements were made after removing the container with the glass particles from the phosphate solution, washing the particles three times with deionized water, and drying for more than 12 h at 90°C. Changes in the pH of the phosphate solution during the conversion reaction were monitored using a pH meter (Accumet-AR25; Fisher Scientific, St Louis, MO).

Structural and chemical characterization

Structural and compositional changes in the glasses which resulted from the reaction with the phosphate solution were monitored using several techniques. Scanning electron microscopy (SEM) and energy-dispersive X-ray analysis (EDS) in the SEM (Hitachi: S-4700) were used to investigate structural and compositional changes of the solid reaction product. Crystalline phases were detected by X-ray diffraction (XRD; Scintag 2000) using Cu K_α radiation (λ = 0.15406 nm) in a step-scan mode (0.05° per step) in the range of 10–80° 2θ. Compositional analysis of the starting glasses and the reaction products was performed using X-ray fluorescence (XRF; XEPOS, Spectro Analytic, Marbough, MA). Fourier transform infrared (FTIR) analysis (Perkin Elmer 1760-X) was performed in the wavenumber range of 400–4000 cm^{-1} on disks prepared from a mixture of 2 mg of the reaction product with 150 mg of high-purity KBr.

Changes in the ionic concentrations of the phosphate solution, resulting from the conversion of the glasses to HA, were also measured. After a given time of reaction between the glass and the phosphate solution, 1 cm^3 of the solution was removed and diluted with deionized water for analysis. The concentrations of Na^+ and Ca^{2+} ions in the solution were measured using atomic absorption (AA) spectroscopy (Perkin Elmer Model 3100). Ion chromatography (IC) was used to measure the PO_4^{3-} concentration, using 2.7 mM Na_2CO_3 and 0.3 mM $NaHCO_3$ as fluent through the column of the Dionex IonPac (AS12A; 4 mm) at a flow rate of 1.5 cm^3/min with the isocratic program. The B^{3+} concentration in the final solutions, after the conversion reaction had effectively stopped, was measured using inductively-coupled plasma mass spectrometry, ICP-MS (ACME Labs, Vancouver, BC, Canada). The Si^{4+} concentration was measured in a photo-spectrometer (Model DR/2010C, Hach Co., Loveland, CO), using a silicomolybdate method (Method 8185) described by the instrument manufacturer.

RESULTS AND DISCUSSION

Figure 1 shows data for the fractional weight loss, ΔW, versus reaction time, *t*, for particles of the four glasses during conversion to HA [$\Delta W = (W_o - W)/W_o$, where W_o is the initial mass and W is the mass at time *t*.] For each glass, ΔW increased with time, eventually reaching a final limiting value, ΔW_{max}. The reaction rate increased with increasing B_2O_3 content of the glass. The difference in reaction rate between the silicate 45S5 (0B) glass and the borate (3B) glass was quite remarkable, with ΔW reaching its maximum value in <60 h for 3B, compared with >600 h for 0B. Figure 2 shows data for the pH of the phosphate solution as a function of reaction time. With increasing B_2O_3 content of the glass, the pH of the solution increased more rapidly with time, reaching a final limiting value at a shorter time, but the final limiting pH value was lower. For the 0B glass sample, a final pH value of 11.5 was reached after ~500 h, whereas it took only ~50 h to reach the final pH of 9.6 for the 3B glass. The data showed that quite high pH values can be reached when silicate and borate bioactive glasses are converted to HA in a static or intermittently-stirred phosphate solution.

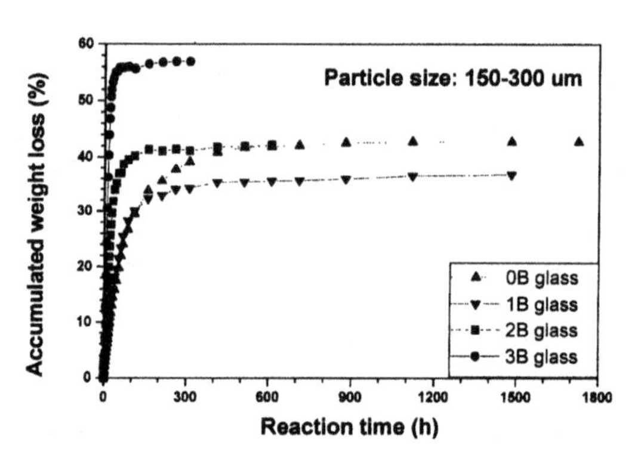

Figure 1. Weight loss versus time during the conversion of silicate (0B), borate (3B), and borosilicate (1B, 2B) glass particles to HA in 0.02 M K_2HPO_4 solution at 37°C.

Figure 1 also shows that the maximum measured weight loss ΔW_{max} was dependent on the B_2O_3 content of the glass. With decreasing B_2O_3 content of the borate and borosilicate glasses, ΔW_{max} decreased from 60% (3B) to 40% (2B), to 35% (1B). However, this decreasing trend was not followed by the silicate glass (0B), for which ΔW_{max} was 40%. It is likely that additional factors contributed to the weight loss of the 0B glass [22]. As found by Iler [23], for pH values above ~9, the solubility of SiO_2 near room temperature increased rapidly with pH. Enhanced SiO_2 dissolution due to the higher pH of the solution might provide an additional factor contributing to the weight loss of the 0B glass.

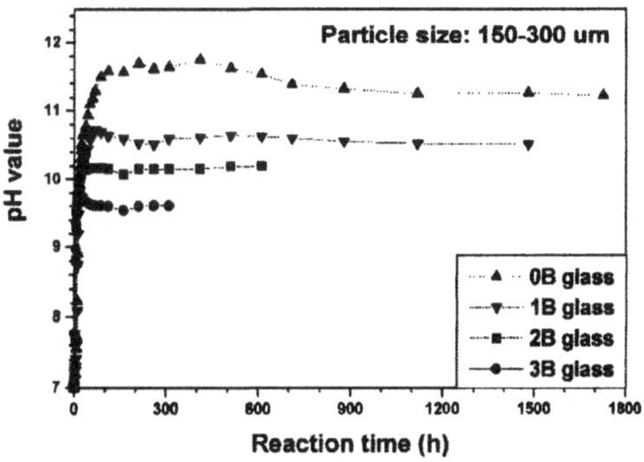

Figure 2. pH versus time during the conversion of silicate (0B), borate (3B), and borosilicate (1B, 2B) glass particles to HA in 0.02 M K_2HPO_4 solution at 37°C.

According to XRD (Fig. 3), the crystalline reaction products of all four glasses consisted of HA. Assuming that the Ca in the glass reacted with the PO_4^{3-} ions from the solution and with P^{5+} ions dissolved from the original glass to form stoichiometric HA, and that the other components of the glass dissolved completely into the solution, then the theoretical weight loss, ΔW_{th}, is calculated to be 56, 57, 58, and 59 wt% for the 0B, 1B, 2B, and 3B glasses, respectively. Figure 1 indicates that for the borate (3B) glass, the measured limiting weight loss, ΔW_{max} = 57 wt%, was close to ΔW_{th}, indicating almost full conversion to HA. However, for the 45S5 (0B), 1B, and 2B glasses, ΔW_{max} was well below ΔW_{th}, indicating incomplete conversion to HA.

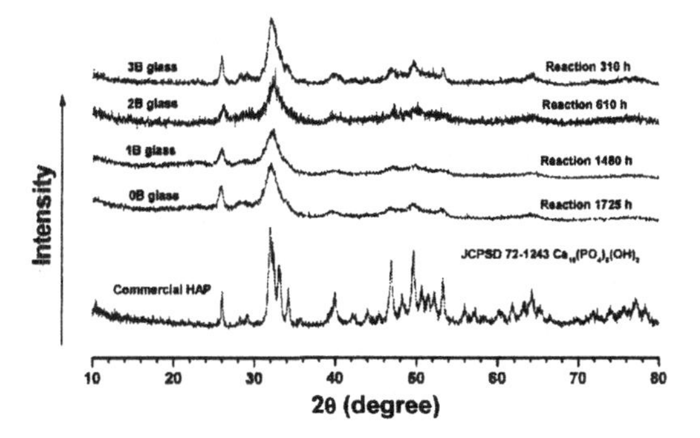

Figure 3. XRD patterns of the as-formed solid products, after reacting 0B, 1B, 2B, and 3B glass particles in 0.02 M K_2HPO_4 solution at 37°C. The pattern of a standard HA is also shown.

The extent of the glass conversion to HA determined from the weight loss data was well supported by chemical analysis. The XRF and EDS data in Table 2 show that CaO and P_2O_5 were the only major components in the reaction product for the 3B glass, with a Ca/P atomic ratio (1.7) almost identical to the value of 1.67 for stoichiometric HA. For the 0B, 1B, and 2B glasses, in addition to CaO and P_2O_5, SiO_2 is a major component of the reaction product. For these glasses, the EDS and XRF data indicated incomplete conversion to HA, with the product consisting of SiO_2-rich core surrounded by HA layer.

Table 2. XRF and EDS data for the composition (in wt %) and Ca/P atomic ratio of the reaction products resulting from conversion of the four glasses to HA in a 0.02 M K_2HPO_4 solution.

Glass	XRF				EDS			
	SiO_2	CaO	P_2O_5	Ca/P	SiO_2	CaO	P_2O_5	Ca/P
0B	30.0	45.2	24.8	2.3	47.3	32.8	19.9	2.1
1B	38.3	41.0	20.7	2.5	34.6	40.9	24.5	2.1
2B	25.6	48.1	26.3	2.3	37.5	36.2	26.4	1.8
3B	0	57.3	42.9	1.7	0	57.3	42.7	1.7

Figure 4 shows FTIR spectra of the final products for the four glasses after reaction in the phosphate solution, along with the spectrum for a commercial HA powder. Whereas the spectra for all four glasses showed the major resonances associated with HA, at wavenumbers of 560 and 605 cm^{-1} characteristic of the P–O resonance in PO_4^{3-} [24], differences were also apparent. The spectra showed a resonance at 450 cm^{-1}, attributed to the Si–O bending vibration in SiO_4^{4-} [18], for the products of the 0B, 1B, and 2B glasses, indicating the presence of residual SiO_2 in the reaction products of the 0B, 1B, and 2B glasses, but not for the 3B glass. The peak at 870 cm^{-1} was attributed to the P–O–H resonance in HPO_4^{2-} [25], whereas the peaks at 1414, 1550, and 1640 cm^{-1} were attributed to the C–O resonance in CO_3^{2-} [26]. The FTIR spectra therefore indicated the presence of some HPO_4^{2-} and CO_3^{2-} in the HA formed from the glasses. Since the present experiments were carried out in air, it is likely that CO_2 from the atmosphere dissolved into the solution, giving rise to the substitution of CO_3^{2-} into the lattice.

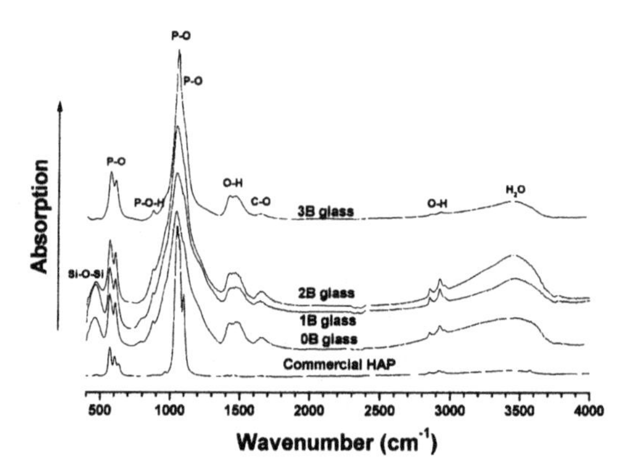

Figure 4. FTIR spectra of the products for 0B, 1B, 2B, and 3B glasses after reaction in 0.02 M K_2HPO_4 solution. For comparison, the spectrum of a commercial HA is also shown.

Figures 5a shows an SEM micrograph of the cross-sections of borate (3B) glass spheres after reaction for 400 hours in the phosphate solution. It should be noted that since the diameters of the spheres (800–1000 μm) were larger than the particle sizes (150–300 μm) used in the work described so far, the fraction of the glass converted to HA after an equivalent time was lower. The cross-sections showed a "composite sphere" microstructure, consisting of a core surrounded by concentric layers. Several cracks, predominantly in the radial direction, presumably caused by capillary stresses during drying, were also present. The cross section consisted of a core (C). a surface layer (S), and at least one intermediate layer (I). EDS analysis indicated that the outer layer S contained no measurable amount of B, and the only elements present were Ca and P, with a Ca/P atomic ratio of 1.65, almost equal to the value (1.67) for stoichiometric HA. The outer layer was therefore fully converted to HA. SEM showed that the HA consisted of a porous structure of nanoscale crystals (Fig. 5b). The elements Ca, P, and B were present in the layer I, as well as in the core C, but the concentration of B was higher in the core. The layer I was apparently an intermediate layer in the conversion process, in which some of the B dissolved into the solution but conversion to HA was not completed. The presence of Na was not detected in any of the layers, indicating that all the Na had dissolved into the solution.

The cross-section of a silicate 45S5 (0B) glass fiber showed that only a thin surface layer had reacted (Fig. 6a). Three structural regions are observed: a surface layer D, an intermediate layer E, and the core F (Fig. 6b). EDS analysis showed that the surface layer contained mainly Ca and P, with a Ca/P atomic ratio equal to 1.6, which is close to the value of 1.67 for stoichiometric HA. The outer layer was therefore HA, formed by the reaction of the glass with the solution. The intermediate layer E, with a thickness of ~2 μm, had a SiO_2 composition predominantly, with small concentrations of Na, K, P, and Ca presumably due to sampling of neighboring layers by the EDS beam and to residual ions not removed completely by washing. The core F had a composition consisting predominantly of SiO_2, CaO, and Na_2O, which was close to that of the 0B glass (Table 1), indicating that the core consisted essentially of unreacted 0B glass.

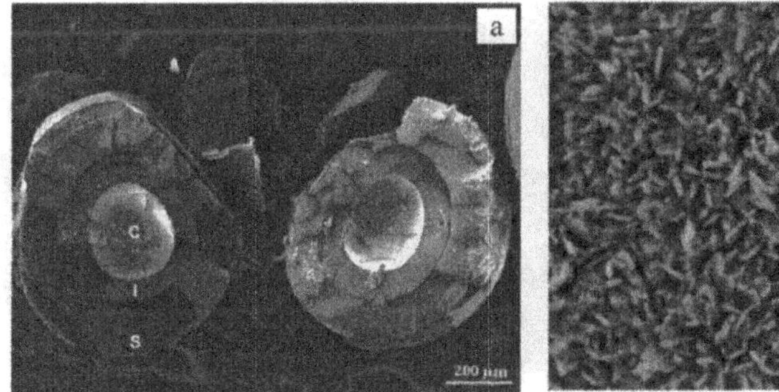

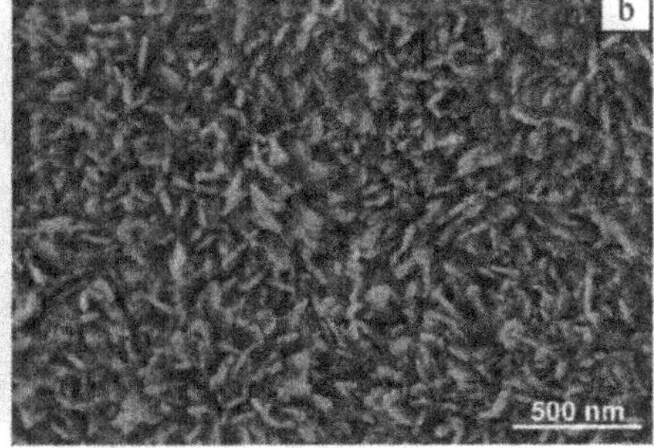

Figure 5. SEM micrographs of (a) the cross-section of borate (3B) glass spheres and (b) the surface of the sphere, after reaction for 400 h in a dilute phosphate solution.

For all four glasses, the Na^+ concentration in the phosphate solution increased with time (Fig. 7), due to Na dissolution from the glass, and eventually reached a final limiting value. On the other hand, the PO_4^{3-} concentration decreased with time, due to reaction with Ca^{2+} to form HA. The initial rate of increase of the Na^+ concentration and the decrease in the PO_4^{3-} concentration were dependent on the composition of the glass, being more rapid for glasses with higher B_2O_3 content. For all four glasses, the Ca^{2+} concentration in the solution throughout the course of the reaction was below the detection limit (<0.5 ppm) of the AA spectrometer. These data are in agreement with the observations of Richard [18], who found that K_2HPO_4 solutions with a wide range of concentrations (0.001–1 M) contained little Ca^{2+} (less than 1–360 ppm) after reacting with the 0B or 3B glass.

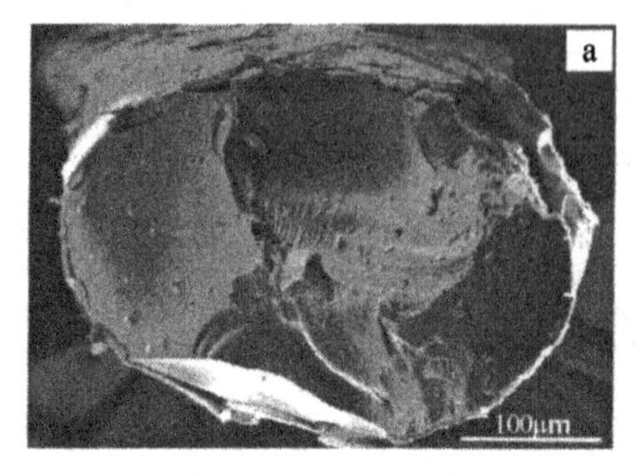

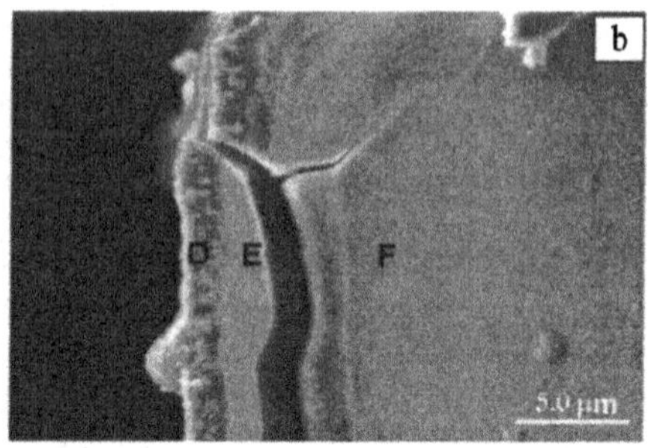

Figure 6. (a) SEM of the cross-section of silicate 0B glass fiber reacted for 400 h in 0.02 M K_2HPO_4 solution at 37°C. (b) Higher magnification SEM micrograph showing a surface layer D, and intermediate layer E, and a core F.

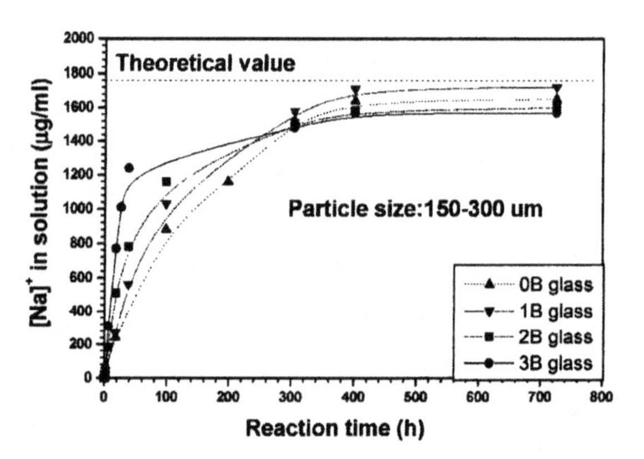

Figure 7. Sodium ion concentration versus reaction time for the solutions resulting from the conversion reactions with the 0B, 1B, 2B, and 3B glass particles.

The measured Na and B concentrations in the final solutions were approximately equal to the theoretical concentrations found by assuming that all the Na and B in the glass dissolved completely into the phosphate solution (Table 3). On the other hand, the measured Si concentration were lower than those calculated from the starting compositions of the glasses, indicating that some Si remained in the solid product. This agrees with the XRF and FTIR data which showed a significant concentration of SiO_2 in the solid product.

Table 3. Measured and calculated ionic concentrations (ppm) in the final solutions.

Glass	Measured					Calculated				
	Si	P	Ca	Na	B	Si	P	Ca	Na	B
0B	1287	140	<1	1650	0	2140	67	0	1820	0
1B	229	160	<1	1720	675	1360	80	0	1780	530
2B	129	137	<1	1600	1154	670	93	0	1740	1030
3B	0	127	<1	1570	1620	0	101	0	1700	1510

CONCLUSIONS

Replacing the SiO_2 in silicate-based 45S5 bioactive glass with varying amounts of B_2O_3 produced borosilicate and borate glasses with controllable conversion rates to HA in dilute (0.02 M) phosphate solution at 37°C. Higher B_2O_3 content of the glass produced a more rapid conversion to HA and a lower pH value of the phosphate solution. Particles of a borate glass (150-300 μm) were fully converted to HA whereas the silicate and borosilicate glasses were only partially converted to HA, resulting in a SiO_2-rich core surrounded by a HA layer. During the conversion reaction, all the Na and B present in the glass particles dissolved completely into the solution. The Ca in the glass reacted with the phosphate ions in the solution to form HA or remained in the unconverted core. The conversion of the borate glass to HA followed a mechanism similar to that for 45S5 glass but without the formation of a SiO_2-rich gel layer.

Acknowledgements: The authors would like to thank Dr. L. Wang (MO-SCI Corp.) for XRF analysis, Dr. K. Kittiratanapiboon and Dr. H. Shi for AA and IC analysis, and Mr. Y. Li for assistance with SEM.

REFERENCES

[1]L. L. Hench, R. J. Splinter, W. C. Allen, and T. K.Greenlee Jr., "Bonding Mechanisms at the Interface of Ceramic Prosthetic Materials," *J. Biomed. Mater. Res.*, **2**, 117-141 (1971).

[2]L. L. Hench and J. Wilson, "Surface-Active Biomaterials," *Science*, **226**, 630-636 (1984).

[3]L. L. Hench, "Bioceramics: From Concept to Clinic," *J. Am. Ceram. Soc.*, **74**, 1487-1510 (1991).

[4]L. L. Hench, "Bioceramics," *J. Am. Ceram. Soc.*, **81**, 1705-1728 (1998).

[5]L. L. Hench, "Bioactive Glasses and Glass Ceramics: A Perspective," pp. 7-23 in *Handbook of Bioactive Ceramics, Vol. 1*, T. Yamamuro, L. L. Hench, and J. Wilson, Eds., CRC Press, Boca Raton, FL, 1990.

[6]T. Nakamura, T. Yamamuro, S. Higashi, T. Kokubo, ans S. Ito, "A New Glass-Ceramic for Bone Replacement: Evaluation of its Bonding to Bone Tissue," *J. Biomed. Mater. Res.*, **19**, 685-698 (1985).

[7]U. Gross and V. Strunz, "The Interface of Various Glasses and Glass-Ceramics with a Bony Implantation Bed," *J. Biomed. Mater. Res.*, **19**, 251-271 (1985).

[8]W. Hoeland, W. Vogel, K. Naumann, and J. Gummel, "Interface Reactions between Machinable Bioactive Glass-Ceramics and Bone," *J. Biomed. Mater. Res.*, **19**, 303-312 (1985).

[9]T. Kokubo, S. Sakka, and T. Yamamuro, "Formation of a High-Strength Bioactive Glass-Ceramic in the System $MgO-CaO-SiO_2-P_2O_5$," *J. Mater. Sci.*, **21**, 536-540 (1986).

[10]S. Yoshii, Y. Kautani, T. Yamamuro, et al., "Strength of Bonding between A-W Glass-Ceramic and the Surface of Bone Cortex," *J. Biomed. Mater. Res.*, **22**, 327-338 (1988).

[11]T. Kitsugi, T. Yamamuro, and T. Kokubo, "Bonding Behavior of a Glass-Ceramic Containing Apatite and Wollastonite in Segmental Replacement of the Rabbit Tibia under Load-Bearing Conditions," *J. Bone Joint Surg.*, **71**, 264-272 (1989).

[12]L. L. Hench and H.A. Paschall, "Direct Chemical Bonding between Bioactive Glass-Ceramic Materials and Bone," *J. Biomed. Mater. Res. Symp.*, **4**, 25-42 (1973).

[13]T. Kokubo, S. Ito, Z. T. Huang, et al., "Ca-P-rich Layer Formed on High Strength Bioactive Glass-Ceramic A-W," *J. Biomed. Mater. Res.*, **24**, 331-343 (1990).

[14]P. Ducheyne, "Bioceramics: Material Characteristics versus *in vivo* Behavior," *J. Biomed. Mater. Res.*, **21**, 219-236 (1987).

[15]S. D. Conzone, R. F. Brown, D. E. Day, and G. J. Ehrhardt, "*In vitro* and *in vivo* Dissolution Behavior of a Dysprosium Lithium Borate Glass Designed for the Radiation Synovectomy Treatment of Rheumatoid Arthritis," *J. Biomed. Mater. Res.*, **60**, 260-268 (2002).

[16]D. E. Day, J. E. White, R. F. Brown, and K. D. McMenamin, "Transformation of Borate Glasses into Biologically Useful Materials," *Glass Technol.*, **44**, 75-81 (2003).

[17]Q. Wang, W. Huang, D. Wang, et al., "Preparation of Hollow Hydroxyapatite Microspheres," *J. Mater. Sci.: Mater. Med.*, 2006; in press.

[18]M. N. C. Richard, "Bioactive Behavior of a Borate Glass," M.S. Thesis, University of Missouri-Rolla, 2000.

[19]J. A. Wojick, "Hydroxyapatite Formation on a Silicate and Borate Glass," M.S. Thesis, University of Missouri-Rolla, 1999.

[20]N. W. Marion, W. Liang, G. Reilly, et al., "Borate Glass Supports the *in vitro* Osteogenic Differentiation of Human Mesenchymal Stem Cells," Mech. Adv. Mater. Struct., **12**, 239-246 (2005).

[21]M. N. Rahaman, W. Liang, D. E. Day, et al., "Preparation and Bioactive Characteristics of Porous Borate Glass Substrates," *Cer. Eng. Sci. Proc.*, **26** [6], 3-10 (2005).

[22]W. Huang, D. E. Day, K. Kittiratanapiboon, and M. N. Rahaman, "Kinetics and Mechanisms of the Conversion of Silicate (45S5), Borate, and Borosilicate Glasses to Hydroxyapatite in Dilute Phosphate Solutions," *J. Mater. Sci.: Mater. Med.*, (2006); in press.

[23]R. K. Iler, *The Chemistry of Silica*, Wiley, New York, 1979.

[24]G. Bordas and C. C. Trapalis, "Fourier Transform and Multi-Dimensional EPR Spectroscopy for the Characterization of Hydroxyapatite Gel," *J. Sol-Gel Sci. Technol.*, **9**, 305-309 (1997).

[25]S. Matsuya and Y. Matsuya, "Effect of Fluoride on Apatite Formation from $Ca_4(PO_4)_2O$ in 0.1 mol/L KH_2PO_4," *J. Mater. Sci.: Mater. Med.*, **9**, 325-331 (1998).

[26]H. Morgan, R. M. Wilson, J. C. Elliott, et al., "Preparation and Characterization of Monoclinic Hydroxyapatite and its Precipitated Carbonate Apatite Intermediate," *Biomaterials* **21**, 617-627 (2000).

Dental Ceramics

VARIABLE FREQUENCY MICROWAVE (VFM) PROCESSING: A NEW TOOL TO CRYSTALLIZE LITHIUM DISILICATE GLASS

Morsi Mahmoud, Diane Folz, Carlos Suchicital and David Clark
Department of Materials Science and Engineering
Virginia Polytechnic Institute and State University
Blacksburg, VA 24061

Zak Fathi
Lambda Technologies
Morrisville, NC 27560

ABSTRACT

The goal of this work is to use variable frequency microwave processing to crystallize lithium disilicate glass ($Li_2Si_2O_5$) into a glass-ceramic material. Variable frequency microwave (VFM) processing significantly reduces the time required for crystallizing lithium disilicate glass into a glass-ceramic in comparison to conventional processing. Moreover, VFM permits more uniform heating and precise control of microwave energy over fixed frequency techniques. The $Li_2Si_2O_5$ glass system provides the basis for a large number of commercially successful glass-ceramic products, such as cookware, radomes, ceramic composites, stovetops and dental crowns. A new generation of dental crowns based on this composition provides a driving force for developing more efficient processing methods.

INTRODUCTION

Microwaves are defined as electromagnetic waves in the frequency range from 300 MHz to 300 GHz. During the Second World War, microwaves began to develop for use in navigation and radar-target detection. The use of microwaves on industrial and domestic applications has increased dramatically in the past 30 years[1]. Microwave processing is an attractive alternative to conventional heating methods. It provides benefits which are not easily obtainable otherwise. It has been used in the processing many materials such as rubber, polymers, ceramics, composites, minerals, soils, wastes, chemicals and powders[2]. Nowadays, the use of microwave processing has extended to industrial and medical sectors[3, 4].

The absorption of microwave energy within the material depends on the incident microwave frequency, the dielectric constant of the material, dielectric loss of the material and the distribution of the electric field within the material[1, 5-7]. In many cases, materials processing using microwave technology have numerous advantages over traditional techniques[8-11]; however, stand alone fixed frequency microwave processing technology suffers from hot spot problems that could influence product consistency[12]. Variable frequency microwave (VFM) processing is an emerging technology that overcomes the inherent difficulties of hot spots experienced in fixed frequency microwave processing and was developed for manufacturing advanced materials[13-15]. By sweeping through a bandwidth of frequencies, the VFM method creates multiple hot spots within the processing cavity that lead to time-averaged uniform heating. The name VFM is derived from the variation of the source frequency over time[14]. There are four controllable parameters that characterize VFM processing: central frequency, bandwidth, sweep-rate and forward power[12]. The central frequency can be adjusted to control the coupling efficiency with the material being processed. The combination of bandwidth and sweep rate around the selected central frequency provides the necessary distribution of microwave energy to carry out uniform heating. The microwave forward power determines the heating ramp rate and can be varied depending on the cure profile[16].

Processing of glass-ceramics is carried out by controlled crystallization of a specially formulated base glass[17]. Initially, a parent glass is formed then heat-treated at a specific temperature to encourage the formation of small and numerous nuclei. These nuclei are allowed to grow at high temperature until perhaps as much as 90% of the sample has crystallized[18]. Glass-ceramic materials offer the possibility of combining the special properties of traditionally sintered ceramics with the unique characteristics of glasses. Moreover, it is possible to develop new glass-ceramic materials with unique characteristics that can not be achieved by other means[19, 20]. The properties of these materials are superior to those of the majority of conventional glass or ceramic materials[20-22].

EXPERIMENTAL WORK

Lithium disilicate glass frit (325 mesh; Specialty Glass, Inc.) was melted in a covered platinum crucible in an electric furnace at 1400°C for 4 hours. The glass samples were cast as rods in a graphite mold (~60 mm long and ~15 mm diameter) at room temperature. Annealing of the glass samples was performed at 400°C for 24 hours in an electric furnace after which they were oven cooled to room temperature.

Glass specimens were cut into ~ 1 cm thick disks using a low-speed diamond saw. Those glass disks were nucleated conventionally in an electric furnace at 485°C for 2 hours. The nucleated glass disks were crystallized using conventional or VFM heating. Conventional crystallization was carried out in an electric furnace at 680°C for 2 hours. Microwave crystallization of the nucleated disks was performed using a MicroCure-Model 2100 (500W; Lambda Technologies). A central frequency of 6.425 GHz with a band width of 1.15 GHz and sweep time of 0.1 sec was used to crystallize the glass samples. The temperature was monitored using an infrared (IR) sensor connected to a RayTeck T30 temperature controller for IR monitoring and closed loop feedback.

Samples of glass, nucleated glass and glass-ceramic were ground and polished for optical microscopy and fourier transform infrared spectroscopy (FTIR) reflectance techniques. The samples were first ground with 180 grit silicon carbide (SiC) paper to remove surface crystallization and to ensure complete flatness of the surface, then polished gradually starting from 600 grit SiC paper and finally ending with 0.25μ diamond compound on a nylon cloth. Infrared spectra were recorded on a Nicolet FTIR instrument (Avatar 330) using a smart stage (SpeculATR) in specular reflectance mode. A 45° angle of incidence beam technique and 7 mm sample mask in the range of 4000–400 cm^{-1} with 32 scans was used. A cavity perturbation technique was used to measure the complex dielectric constant of the lithium disilicate glass, nucleated glass and glass-ceramic between room temperature and 600°C in the frequency range 400 MHz to 2466 MHz in air. The cavity perturbation measurement technique is based on knowing the difference in the cavity response between an empty sample holder and one containing a sample at each temperature. The measurements were performed at Chalk River Nuclear Labs of Atomic Energy of Canada Limited (AECL labs) by Dr. Ron Hutcheon.

A simultaneous thermal analysis instrument (NETZSCH model STA 449 C Jupiter®) was used to identify the nucleation and crystal growth temperature peaks for the glass. The temperature range was from room temperature to 1100°C in air with a heating rate of 10°C/min in Pt/Rh crucibles. The density of the glass and glass-ceramic samples was measured by means of a pycnometer (Micromeritics -AccuPye ^{TM} 1330) using helium gas.

A mortar and pestle were used to grind the glass and glass-ceramic samples for x-ray analysis. The x-ray patterns were obtained using an x-ray diffraction spectrometer (Scintag, XDS 2000) with a Copper (Cu) tube at 45 kV and 40 mA. A scanning speed of 2 degrees (θ)/min was used for the analyses. The reference data for the interpretation of x-ray diffraction patterns were obtained from the ASTM x-ray diffraction file index.

The microstructures of heat-treated glass samples were examined using a scanning electron microscope. The fractured surfaces of the glass-ceramic samples were first etched for 2 minutes in 2% diluted hydrofluoric acid (HF) acid, rinsed with distilled water and acetone, dried and then coated with a thin film of evaporated carbon in order to overcome the effect of sample charging under the electron beam.

RESULTS AND DISCUSSION

Thermal Analysis

Thermal analysis is a powerful tool used in detecting and analyzing the thermal effects produced during crystallization of glass. Based on the thermal analysis data, the glass transition temperature (T_g) is detected at 430°C followed by an endothermic peak that is characteristic of the nucleation process (heat absorbed to overcome the nucleation reaction) in the glass at a peak temperature of 488°C. The crystallization exothermic peak is detected at 683°C and represents the heat released due to the crystallization reaction. The same crystallization peak has been reported by Soares et al[23] for the bulk stoichiometric composition of lithium disilicate glass. At 956°C and 1033°C two endothermic peaks were detected. Those two endothermic peaks are characteristics of the polymorphous transformation of the stoichiometric lithium disilicate crystal phase at 956°C and of the congruent melting at 1033 °C of the same crystalline phase[21]. Thermal analysis has been used to confirm the chemical composition of the glass by comparing the thermal analysis data with the lithium silicate phase diagram[21]. All the glass disks were conventionally nucleated at 488°C before crystallizing them by VFM or conventional heating.

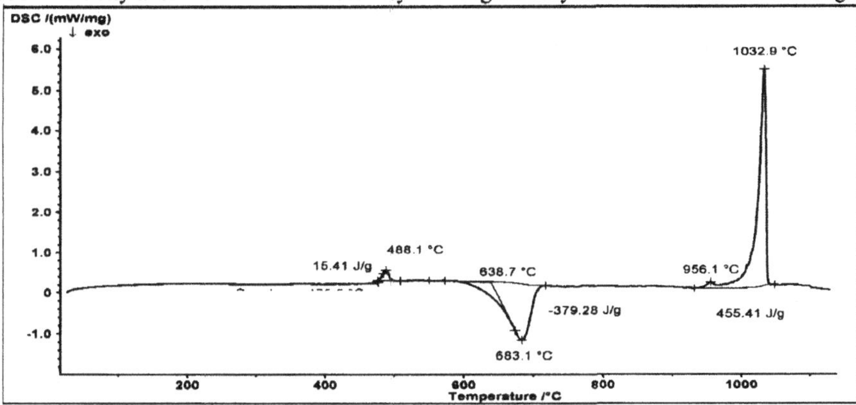

Figure 1. DSC chart of bulk annealed lithium disilicate glass sample.

Density Measurements

The density values for lithium disilicate glass frit, bulk glass, and glass-ceramics crystallized by VFM are shown in Table 1. It was observed that the densities of the glass-ceramics were higher than the corresponding glass. This increase may be attributed to the fact that, in most cases, the densities of the crystals (due to higher structural compaction) are higher than those of the glass of the same composition[24]. The density of the glass-ceramic crystallized by VFM (heat treated at ~ 600°C, 2 minutes) is not only higher than the corresponding glass-ceramic crystallized conventionally (heat treated at 680°C,120 minutes), but also very close to the theoretical density of lithium disilicate crystals. That fact implies that in a significantly shorter time a very dense microstructure with almost no pores resulted from VFM heating as compared to glass-ceramic produced using conventional heating.

Table 1. Density measurements of $Li_2Si_2O_5$ glass frit, bulk glass, and glass-ceramic samples.

Density measurement	Glass frit (Powder)	Bulk glass	Glass-ceramic (Conven.)	Glass-ceramic (VFM)	Theoretical density of $Li_2Si_2O_5$ crystals[25]
Average density (g/cm^3)	2.3401	2.3481	2.4281	2.4383	2.439
Standard deviation	0.0004	0.0001	0.0005	0.0010	

Fourier transform infrared spectroscopy (FTIR)

Figure 2 provides IR spectra for annealed lithium disilicate glass and for the nucleated glass sample. The figure shows that the nucleated glass has the same IR pattern as the annealed glass indicating that the IR technique is not sensitive to the nuclei formation in the glass structure and also indicating that the nucleated sample did not crystallized yet.

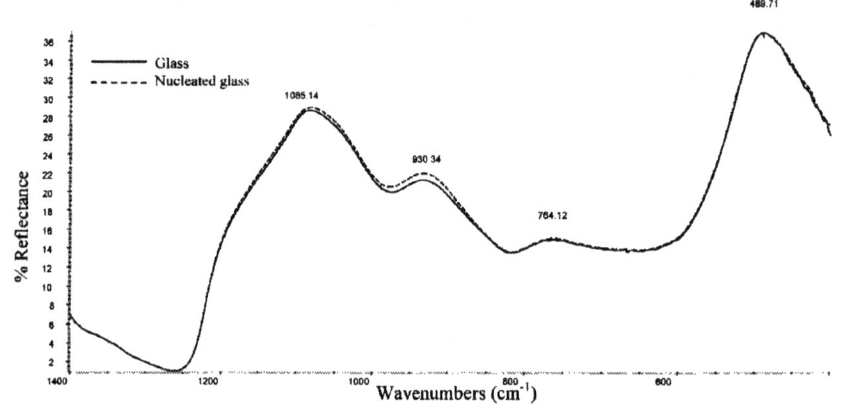

Figure 2. IR spectra of lithium disilicate glass and nucleated glass.

The strong band at 1085 cm^{-1} in the spectra of silicates glass is known to be due to (i) the stretching mode of the Si-O- terminal group (main contribution to that band) and (ii) the highest frequency component of the asymmetric stretching mode of the Si-O-Si[26, 27]. The band at 930 cm^{-1} is known to be due to the principle (lower frequency) component of the asymmetric stretching mode of the Si-O- and (-O-Si-O-)[28, 29]. The band at 764 cm^{-1} is due to the symmetric stretching of the Si-O-Si bond ,while the band at 490 cm^{-1} can be assigned to the bending vibration of the Si-O-Si bond[30].

Figure 3 shows the IR spectra of glass-ceramics crystallized conventionally at 680°C for 2 hours and by VFM processing at ~ 600°C for 2 minutes. As can be seen from the figure, not only did both samples have almost the same peak positions, but the VFM crystallized sample had a slightly higher intensity than the conventional one. The assignment of the IR bands is given in Table 2. In summary, both samples developed the same structure in dramatically different times, indicating that the kinetics of crystallization in the VFM field are different and faster than those experienced conventional during heat treatment.

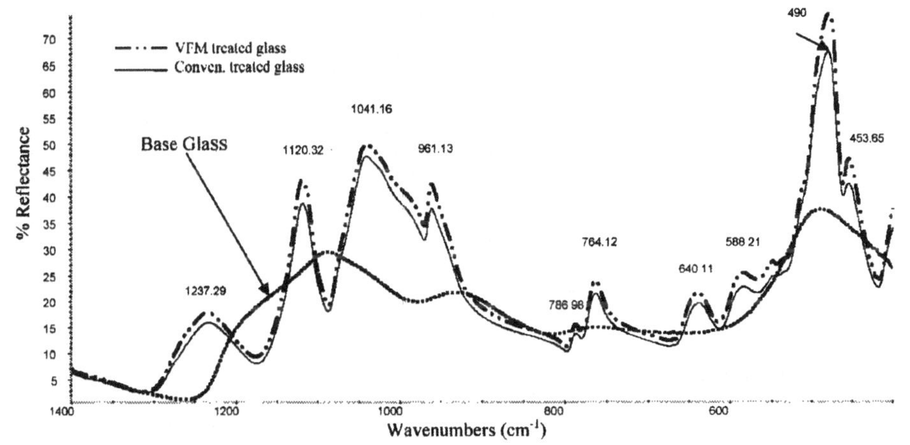

Figure 3. IR spectra of two glass-ceramics samples crystallized by VFM (~ 600°C for 2 minutes) and conventionally (680°C for 2 hours). The base glass spectrum is shown for comparison.

Table 2. IR bands for lithium disilicate glass-ceramics.

IR bands (cm^{-1})	Assignment of the IR bands	IR bands (cm^{-1})	Assignment of the IR bands
1237	$Si-O^-$ & (Si-O-Si) asym.stretching	640	(Si-O-Si) sym. stretching
1120	$Si-O^-$ & (Si-O-Si) asym. stretching	584	(Si-O-Si) sym. stretching
1041	$Si-O^-$ & (Si-O-Si) asym.stretching	550	(O-Si-O) bending mode
961	(Si-O-Si) asym. stretching & $Si-O^-$	480	(O-Si-O) bending mode
787	(Si-O-Si) sym. stretching	454	(O-Si-O) bending mode
764	(Si-O-Si) sym. stretching		

Complex Dielectric Measurements

The fundamental properties measured by the cavity perturbation technique are the real part of the dielectric constant, represented by ε', and the imaginary or absorptive part of the dielectric constant, represented by ε''. Figure 4 shows ε'' for the $Li_2Si_2O_5$ glass and the corresponding glass-ceramic material in the 2.46 GHz range. At room temperature, the dielectric loss factor for the $Li_2Si_2O_5$ glass (0.149) is approximately 8 times higher than that of the corresponding glass-ceramic material (0.019). Thus the microwave absorption at room temperature in the glassy phase is higher than that in the crystalline phase based on the following equation (1):

$$P_a = \omega\, \varepsilon_0\, \varepsilon''_{eff}\, E^2_{rms} \tag{1}$$

Where ω= angular frequency = $2\pi f$ (f = operating frequency in hertz);

ε_0= permittivity of free space = 8.85×10^{-12} farads/m;

ε''_{eff}= effective relative dielectric loss (dissipation) factor (unitless); and,

E_{rms}= root mean square of the internal electric field (volts/m).

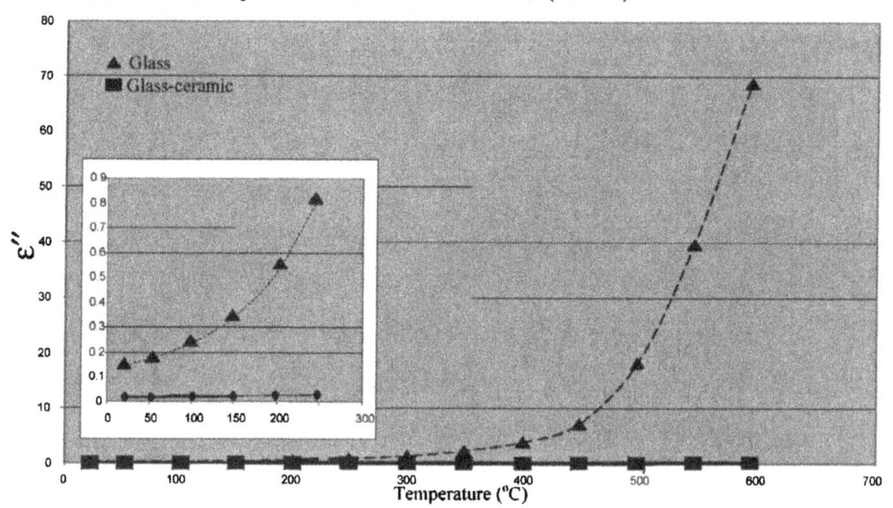

Figure 4. Dielectric loss measurements of lithium disilicate glass and glass-ceramic at 2.46 GHz.

This dielectric measurement data is in good agreement with what has been observed in VFM crystallization experiments. The power absorbed by the sample (rise of the sample temperature) at the beginning of the VFM crystallization experiment (glass sample) was higher than that of the same sample after it was crystallized (2 minutes at ~600°C). After crystallization, the sample temperature did not rise even under higher applied forward power (Figure 5). The change in the dielectric properties can be attributed to the change from amorphous (open) structure of the glassy phase (higher dielectric loss), which is heated more easily by microwave energy, to the well ordered (compact) structure of the corresponding glass-ceramic phase (low dielectric loss) that is harder to heat with microwave energy under normal conditions[4, 31]. The considerable change in dielectric properties of $Li_2Si_2O_5$ glass after crystallization can be used as a signal or a sensor for the completion of the microwave crystallization process.

Scanning Electron Microscopy

The scanning electron micrograph of the as fractured surface of a glass-ceramic sample crystallized by VFM is shown in Figure 6. The glass-ceramic sample exhibits the tightly interlocking tabular crystal form that is characteristic of the corrugated sheets or layers of lithium disilicate glass-ceramics. As a result of this microstructure, good isotropic mechanical properties are achieved[21, 32]. Thus the higher the degree of interlocking between crystals (high crystalline content), the higher the mechanical strength and toughness of the lithium disilicate glass-ceramic[21].

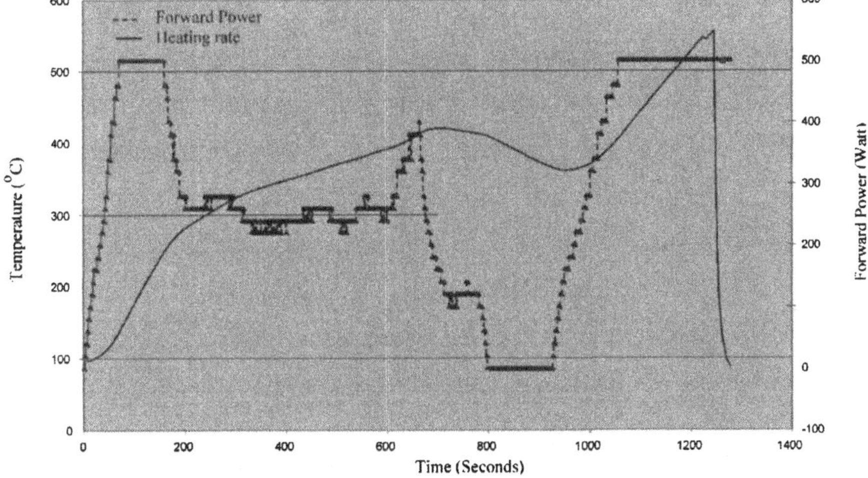

Figure 5. Heating rate and forward power during VFM crystallization of the glass at 6.425 GHz.

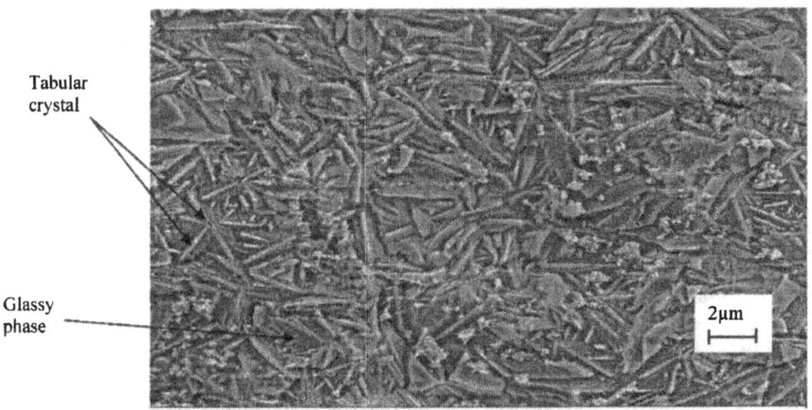

Figure 6. Micrograph of fractured surface of VFM crystallized lithium disilicate glass-ceramic.

X-ray Diffraction

Figure 7 shows the x-ray patterns of annealed lithium disilicate glass and glass-ceramic samples crystallized by VFM. The prepared glass was quenched successfully and the amorphous structure was achieved with the characteristic short range order pattern. These samples were crystallized by VFM into lithium disilicate glass-ceramic with its characteristic peaks. The crystal phase in the glass-ceramic was identified as orthorhombic lithium disilicate whose features include corrugated sheets of $(Si_2O_5)^{-2}$ on the (010) plane[21]. Identification of the crystal phase and the crystallographic planes corresponding to the 2θ value for all the peaks was performed using the ASTM x-ray diffraction file index and from several other publications[25, 33-41].

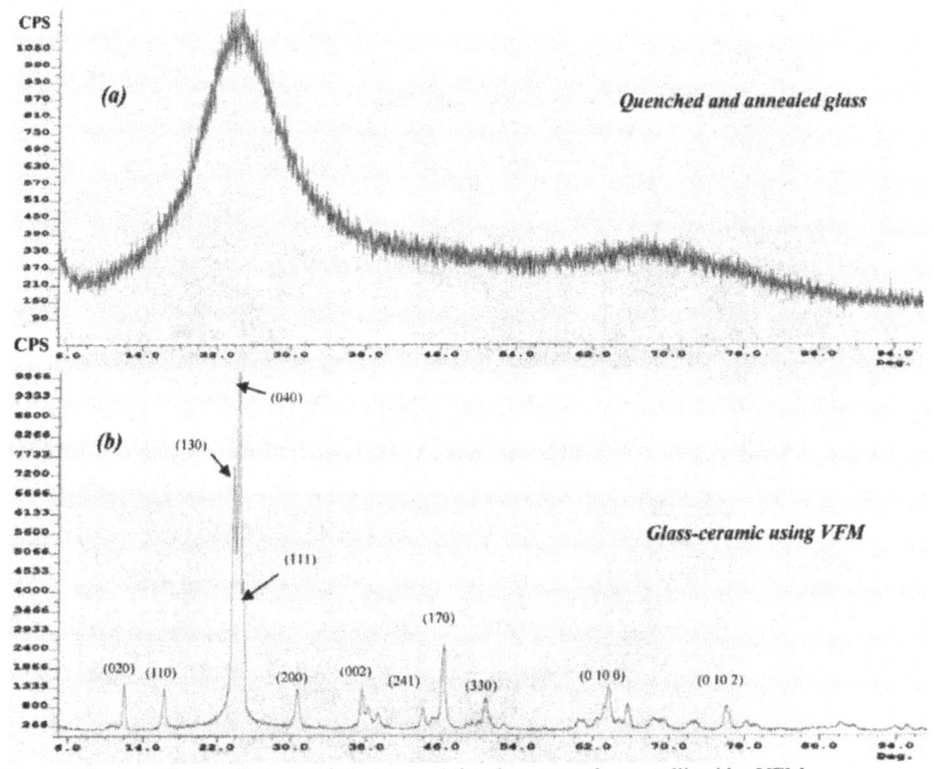

Figure 7. X-ray patterns of (a) annealed glass and (b) glass-ceramic crystallized by VFM.

CONCLUSION

Lithium disilicate glass has been successfully crystallized directly and rapidly using a VFM field in a significantly shorter time when compared to conventional heating. The density of a two minute VFM treated sample is almost the same as the theoretical density of lithium silicate crystals and also slightly higher than the corresponding conventionally treated sample. The FTIR and XRD analyses have confirmed the crystallization of lithium disilicate glass by VFM. Samples crystallized by VFM or conventional heating have been shown to exhibit the same crystal structure and crystalline phase. The dramatic change in the dielectric properties of the glass after crystallization could be used as a sensor for completion of the crystallization process in the microwave field. Microwave processing, and especially VFM, could be implemented in manufacturing operations for the next generation of advanced glass-ceramics products (such as dental crowns) to achieve rapid processing and possibly improved properties and process reliability.

ACKNOWLEDGEMENTS

We would like to thank Dr. Ekkehard Post from NETZSCH Instruments, Inc. for performing the thermal analyses. Also, we would like to acknowledge Dr. Ron Hutcheon from Microwave Properties North (Ontario, Canada) for performing the dielectric measurements.

REFERENCES

1. Metaxas A.C. and R.J. Meredith, *Industrial Microwave Heating*. 1983, London,UK: Peregrinus.
2. *Materials Research Advisory Board,Microwave Processing of Materials* 1994: National Research Council, National Academy Press.
3. Clark, E.D., et al., *Microwave Soultions for Ceramics Engineers*. 2005, Westerville,Ohio: The American Ceramic Society. 494.
4. Clark, D. and D. Folz, *Microwave processing of materials.* Advances in Science and Technology (Faenza, Italy), 2003. **31**(10th International Ceramics Congress, 2002, Part B): p. 367-380.
5. West, J.K. and D.E. Clark, *Microwave absorption by materials: Theory and application.* Ceramic Transactions, 2000. **101**(Surface-Active Processes in Materials): p. 53-73.
6. Folz, D.C., et al., *Microwave and Radio Frequency Applications: Bridging Science, Technology, and Applications. (Proceedings of the Third World Congress held in Sydney, Australia September 2002.)*. 2003. 519 pp.
7. Copson, D., *Microwave Heating: In Freeze Drying, Electronic Ovens and Other Applications*. Avi Publishing Company. 1962, Westport, CN.
8. Clark, D.E., Sutton W. H., and Lewis D.A. *Microwave Processing of Materials.* in *Microwaves:Theroy and Application in Materials Processing IV*. 1997. Westerville,OH: Ceramic Transactions,American Ceramic Society.
9. Clark, D.E. and D.C. Folz. *Developments in microwave processing technologies.* in *Ceramic Engineering and Science Proceedings (USA). Vol. 18, no. 4, pp. 531-541.* 1997.
10. Clark, D.E., et al., *Applications of microwave processing in ceramics and waste remediation.* Ceramic Transactions, 1997. **80**(Microwaves: Theory and Application in Materials Processing IV): p. 507-514.
11. Leiser, K.S. and D.E. Clark, *Microwave behavior of silicon carbide/high alumina cement composites.* Ceramic Transactions, 2001. **111**(Microwaves: Theory and Application in Materials Processing V): p. 267-275.
12. Fathi, Z., et al., *Industrial applications of variable frequency microwave energy in materials processing.* Materials Research Society Symposium Proceedings, 1996. **430**(Microwave Processing of Materials V): p. 21-28.
13. Fathi, Z., et al., *Curing polymer layers on semiconductor substrates using variable-frequency microwave energy.* 1998, (Lambda Technologies, Inc., USA; Fathi, Zakaryae; Tucker, Denise A.; Garard, Richard S.; Wei, Jianghua). Application: WO. p. 33 pp.
14. Garard R. S. , Fathi Z. , and Wei B. *Materials Processing Via Variable Frequency Microwave Irradiation.* in *Microwaves: Theory and Application in Materials Processing III*. 1995. Cincinnate,USA: Ceramic Transactions,The American Ceramic Society.

15. Lauf, R.J., et al., *Uniformly heating substrates using variable frequency microwave energy*. 1998, (Lambda Technologies Inc., USA; Lockheed Martin Energy Research Corp.). Application: US. p. 28 pp , Cont -in-part of U S 5,721,286.
16. Anderson, B., et al. *Rapid Processing and Properties Evaluation of Flip-Chip Underfills*. in *Proc., Technical Program NEPCON West '98 Conference,*. 1998. Anaheim, CA.
17. Barsoum, M.W., *Fundamentals of ceramics*. 1997: McGraw-Hill International editions,p. 322.
18. Berezhnoi, A.I., *Glass-ceramic and photositalls*. 1970, New York Plenum press,P.2.
19. James, P.F.J., R.W, *Glass-ceramics in "High performance glasses"*, ed. M. Cable and J.M. Parker. 1992, New York: Chapman & Hall,P.102.
20. Mc Millan, P.W., *Glass-ceramics*. 1979, London: Academic press,2nd edition,P.1.
21. Holand, W. and G. Beall, *Glass-ceramic technology*. 2002: The American Ceramic Society.
22. Hench, L.H. *Structure and properties of glass-ceramics*. in *4th space congr., Cocoa Beach Florida*. 1967.
23. Soares Jr, P.C., et al., *TEM and XRD study of early crystallization of lithium disilicate glasses*. Journal of Non Crystalline Solids, 2003. **331**: p. 1-3.
24. Cumpston, B., F. Shadman, and S. Risbud, *Utilization of coal-ash minerals for technological ceramics*. J. Mater. Sci., 1992. **27**: p. 1781-84.
25. De Jong, B.H.W.S., *2000 JCPDS International Diffraction Data;file 82-2396*. Journal of Non-Crystalline Solids, 1994. **176**: p. 164.
26. Efimov, A.M., et al., *Infrared reflection spectra,optical constants and band parameters of binary silicate and borate glasses obtained from water free polished sample surface*. Glass Technology, 2005. **46**(1): p. 20-27.
27. Fuxi, G., *Optical and Spectroscopic properties of glass*. 1992: Springer-Verlag. 283.
28. Efimov, A.M., *Vibrational spectra, related properties and structure of inorganic glasses*. Journal of Non Crystalline Solids, 1999. **253**: p. 95-118.
29. Fuss, T., *Effect of pressure on crystallization in lithium disilicate glass*, in *Ceramic Engineering*. 2004, University of Missouri-Rolla. p. 84.
30. El-Alaily, N.A., *Study of some properties of lithium silicate glass and glass ceramics containing blast furnace slag*. Glass Technology, 2003. **44**: p. 30-35.
31. A.D.Cozzi, et al. *Nucleation and crystallization of Li2O-2SiO2 in a 2.45 GHz microwave field*. in *17th Annual Conference on Composite and Advanced ceramic Materials*. 1993.
32. Krishnaswami, S., et al., *Connection between the microwave and far infrared conductivity of oxide glasses*. Journal of Non Crystalline Solids, 2000. **274**: p. 307-312.
33. Hesse, K., *JCPDS-International Center for Diffraction Data, file 29-0829*. Acta Crystallographica, 1977. **33**(Structural Science,Section B): p. 901.
34. Ray, C.S., X. Fang, and D.E. Day, *New method for determining the nucleation and crystal-growth rates in glasses*. Journal of the American Ceramic Society, 2000. **83**: p. 865-872.
35. Smith, R.I., *JCPDS-International Center for Diffraction Data, file 80-1470*. Powder Diffraction, 1990. **5**(137).
36. Smith, R.I., *JCPDS-International Center for Diffraction Data, file 79-1899*. Acta Crystallographica, 1990. **46**(Section C): p. 363.
37. Kalinina, A., *JCPDS-International Center for Diffraction Data, file 24-0651*. Inorg. Mat. er. (Engl. Transl.), 1970. **6**: p. 389.

38. Liebau, F., *JCPDS-International Center for Diffraction Data, file 40-0376.* Acta Crystallographica, 1961. **14**: p. 389.
39. West, A.J., *JCPDS-International Center for Diffraction Data, file 30-0767.* Journal of American Ceramic Society, 1976. **59**: p. 124.
40. Maksinov, B.A., *JCPDS-International Center for Diffraction Data, file 74-2145.* Dokl. Akad. Nauk. SSSR, 1968. **178**: p. 1309.
41. Voellenkle, H., *JCPDS-International Center for Diffraction Data, file 63-1517.* Z. Kristallogr, 1981. **77**: p. 154.

Author Index

9 780470 080566